T0094013

Big Data Analytics Strategies for the Smart Grid

Big Data Analytics Strategies for the Smart Grid

Carol L. Stimmel

CRC Press
Taylor & Francis Group
Boca Raton   London   New York

CRC Press is an imprint of the
Taylor & Francis Group, an **informa** business

AN AUERBACH BOOK

CRC Press
Taylor & Francis Group
6000 Broken Sound Parkway NW, Suite 300
Boca Raton, FL 33487-2742

© 2015 by Carol L. Stimmel
CRC Press is an imprint of Taylor & Francis Group, an Informa business

No claim to original U.S. Government works

Printed on acid-free paper
Version Date: 20140611

International Standard Book Number-13: 978-1-4822-1828-2 (Hardback)

Library of Congress Cataloging-in-Publication Data

Stimmel, Carol L., 1965-
    Big data analytics strategies for the smart grid / Carol L. Stimmel.
        pages cm
    Summary: "A successful data analytics program is the only way utilities will be able to meet the challenges of a modernized grid that is responsive from an operational perspective and meets the demands of greenhouse gas legislation. This book addresses the needs of applying big data technologies and approaches, including Big Data cybersecurity, to the critical infrastructure that makes up the electrical utility grid. It explores the unique needs of electrical utility grids, including operational technology, IT, storage, processing, and transformation to a usable state that benefits the utility business and electricity consumers"-- Provided by publisher.
    Includes bibliographical references and index.
    ISBN 978-1-4822-1828-2 (hardback)
    1. Smart power grids. 2. Big data. 3. Electric power distribution. I. Title.

TK3105.S75 2014
621.310285'57--dc23   2014019445

Visit the Taylor & Francis Web site at
http://www.taylorandfrancis.com

and the CRC Press Web site at
http://www.crcpress.com

Dedication

To my beloved son Jake—for your infectious enthusiasm, pursuit of
new ideas, persistent distrust of the status quo, and brilliant
sense of humor. I love you, honey.

Contents

Foreword

The Data Era of Energy

This book provides an in-depth analysis that will help utility executives, as well as regulators, investors, large power users and entrepreneurs, understand some of the tectonic changes coming to an industry that from the outside can seem impervious to change. Making sense of a chaotic future, Carol charts a path where everyone can benefit.

– Dr. Amit Narayan

Predictive analytics and data systems will have a transformative impact on the electricity industry. Not only will the integration of Big Data technologies help make the grid more efficient, it will fundamentally change who sells electric power; how it gets priced; and how regulators, utilities, grid operators, and end-users interact.

To understand the full impact that software and hardware will have, one has to first take a step back and take a look at how vastly different the electric power industry is from most others. Utilities typically don't compete over market share or increasing revenue and margins. Instead, they exist through legal monopolies, enjoy stable pricing, and can recover investments in fixed assets in a predictable way. Instead of revenue, their biggest concern is quality of service, stability, and reliability. They are public or quasi-public institutions with fiduciary responsibilities that provide an essential service to everyone within a service territory, with a vital impact on public safety and well-being. If Facebook shuts down for two hours, Twitter is abuzz with jokes. If a utility experiences a two-hour blackout, executives have to explain what happened to consumer groups and PUC officials. The Department of Energy estimates that blackouts and power quality issues currently cost American businesses more than $100 billion each year.

To achieve these demanding levels of performance, utilities have focused on integrating multiple levels of redundancy and control. Peak power plants cost hundreds of millions of dollars and might only be used 50 hours a year, but utilities build them because they are a proven (if inefficient) tool for counteracting temporal spikes in demand. Some of the objections to renewable sources like solar and wind have been because of the variability that they can introduce. Utilities have compensated for a lack of visibility and predictability through buffering, brute-force engineering, and deliberately circumscribing options for the sake of control and consistency.

An Internet for Energy changes this paradigm by providing utilities with real-time feedback and insight for the first time. Simply put, utilities are going to finally know what their customers are doing and what they want. Today, forecasting is done at the system level and it's a fundamental operation that drives practically all operational and planning decisions at the utility. The ability to forecast every meter, transformer, feeder, and province allows us to improve the quality of these decisions and shave billions in operating expense. Just a 0.1% improvement in forecasting at a mid-size European utility can help reduce about $3 million in operating cost in the imbalance markets.

What will that empower them to do in practice? One of the first major impacts will be in the rapid acceptance of demand response. FERC estimates that demand response systems—which are essentially cloud-based platforms for dynamically controlling power consumption—can replace 188GW of demand and avoid $400 billion in peaker plant investments in the U.S. alone. Demand response, however, has largely been the province of large utilities and large customers because the hardware systems required for conventional demand response have been unaffordable for most.

Software-based demand response reduces the cost of implementing demand response by up to 90 percent. More importantly, it introduces the concept of visibility to demand response. Utilities, or rather, the cloud-based platforms employed by utilities, can look at the consumption patterns of millions of its users at once and rapidly determine which customers would be willing to participate in a DR event, how much these customers will charge for participation, and how much was actually saved.

Demand response shifts from being an expensive technology deployed only a few times a year to a control system that a utility can use on a daily basis for helping consumers save money, meet community emissions standards, and maximize the return on fixed assets like power plants.

Software-based systems, unlike hardware, also improve over time. Think of your own experience with Google, Amazon, or Netflix. Those web platforms improve over time as they absorb and analyze more data. Similarly, software-based demand response systems will become more surgical in how they harvest power. Forecasts can be issued for millions of customers every few minutes to fine-tune predictions for power consumption across an entire region, in specific geographic areas, or users along particular distribution branches. The impact ultimately will be unobtrusive. Consumers and businesses won't know they are saving power until they get a pleasant surprise on their bill.

Similarly, software can supplant traditional hardware systems for frequency regulation and spinning reserves, lowering capital costs while improving performance and accelerating adoption. Like we've seen with the Internet, the applications for systems that can provide granular predictability might as well be infinite.

Greater control and flexibility in power consumption and delivery in turn will pave the way for increased solar, wind, EVs, and storage. These technologies can be integrated safely and more easily when supplemented by digital control system and give their owners a more rapid return on investment. The price gap between peak and off-peak power will begin to erode.

From there, one can see how the underlying business will change. In fact, it is already changing. An estimated 44% of electricity in North America is sold in deregulated markets. Comcast and others are entering home energy and retail space. Electricity retailers in Europe, Australia, and New Zealand (where deregulation has taken place) compete with others to acquire and retain customers and are using ways to micro-segment habits and load profiles.

With dynamic data analysis and control, more power providers will be able to link to the grid and sell power as supply and demand become more fluid and interconnected. Consumers and businesses, likewise, will become more adept at monetizing their consumption patterns. Electricity retailers—already a common feature in Texas and the U.K.—will expand to other markets as deregulation becomes feasible thanks to technology.

Efficiency will also finally take off simply because it will be possible to take actions to curb power easily, as well as measure and monitor the results of efficiency initiatives. Consumers and business, meanwhile, are equally hamstrung. Commercial buildings consume 18% of all of energy in America, but close to 30% of the total is lost through waste or inefficiency, according to the Department of Energy and Environmental Protection Agency. Lighting consumes 19% of all electricity worldwide—more than is produced by nuclear plants and hydroelectric plants combined—but systems that automatically dim lights remain rare. When you look out on a glittering urban skyline at night, you aren't just looking at a scenic vista. You're looking at a tremendous, chronic, and seemingly unstoppable waste.

Similarly, industrial customers will begin to adopt cloud-based systems to help control demand charges. Demand charges can account for 30 percent of a large power user's bill. By employing intelligent automation, large power users can turn down less essential power consumption (like daytime lighting), maintain production flows, and avoid excessive peaks. Without data, large power users can only guess what their power demands might be; data effectively eliminates risk by tightly defining probable outcomes.

Data can also be used to throttle power theft. The World Bank estimates that $85 billion in power gets stolen every year. In emerging nations, the problem is a never-ending crisis: approximately 30 percent of the electricity gets stolen in India, leading to chronic outages, lower productivity, and higher rates. But it's also a problem in the U.S., with $5 billion alone being siphoned off by illegal marijuana growing operations.

The impacts will even be more eye-opening in emerging nations. The International Energy Agency estimates that over 1.2 billion people worldwide do not have access to electricity and over 2.6 billion do not have access to clean cooking facilities. To compound the problem, grid power in many of these countries continues to be supplied by dirty, expensive, and inefficient diesel generators. The root cause of the situation can be traced back to the limitations of the architecture of the supply-centric grid. Microgrids—animated by solar panels, battery banks, and intelligent data systems—will fill this gap.

Granted, the energy-data nexus has had a rocky start. The initial rollout of smart meters—a foundational element of the data era—drew many critics. Customers of Pacific Gas & Electric staged protests against smart meter installations, asserting that the meters posed a health hazard and invaded privacy.

But when you get past the some of the controversial headlines, a different picture emerges. Oklahoma Gas & Electric (OG&E) over the last five years has conducted one of the most successful projects to date for employing data to control energy costs, consumption, and emissions. The utility uses technology from Silver Spring Networks, AutoGrid Systems, and others to deliver information to consumers about peak pricing, manage time-of-user programs, and other initiatives. In an early test with 6,500 customers, the reactions were almost uniformly positive. Customers said they didn't know how peak pricing could lower their bills, and many changed how and when they used air conditioners and washing machines.

OG&E has since expanded the program to 70,000 customers and anticipates growing the number of participants to 120,000 by the end of 2014. OG&E, which also won a J.D. Power Customer Service Award, a rarity for a utility, believes that data systems will help it avoid building any new fossil plants before 2020.

The global proliferation of smart devices will ultimately generate a veritable tidal wave of digitized information. A typical smart meter is serving up 2,880 meter reads a month, versus the one per month delivered by an analog meter. By 2020, the 980 million smart meters worldwide alone will generate 431,000 petabytes of data a year. Building management systems for office buildings will generate around 100 gigabytes of information a year.

Implementing and integrating data systems will take time. Caution and security must still underpin any changes. Still, change is inevitable. How exactly consumers will interact with data remains to be seen, but I don't think the industry will turn back.

Data is the new power.

– Dr. Amit Narayan

Preface

This is a practical book, to be sure, but it is also a book about hope and positive change. I am quite sincere. The delivery of electricity is deeply rooted in the principle of universal access; when clean, reliable energy is available it contributes to poverty alleviation, improved social conditions, and enhanced economic development. In the developed world, we know this to be true. The digital fabric of our lives is a testimony to the importance of energy security. Across the globe, we have seen the vital contributions that electrification has brought to the development of economies and an enhanced quality of life. Nonetheless, this supreme engineering achievement has languished, and we are deeply challenged.

Modern electrification systems are degrading and inefficient in myriad ways, yet the complex and difficult operating conditions of the energy business have been slow to adapt and advance to improve these circumstances. However, with the advent of the information-enabled, two-way grid, we have an opportunity to meet these challenges directly. It is the thesis of this book that through the application of big data analytics and subsequent improvements in situational awareness of the millions of miles of grid across the world, we will be able to integrate renewable energy systems, introduce economic and operating efficiencies, and bring energy services to the over 1 billion people across the world who have no electricity. It is also the view of this work that utilities are confronted with a very difficult charge indeed—to evolve rapidly towards a business standpoint that capitalizes on these key technologies. It is going to take a resolute effort from technologists, utility stakeholders, political bodies, and energy consumers to protect and improve the performance of the grid, as well as affect the change necessary to shield our economies and defend the environment.

I hope to shed light on the considerable potential that big data analytics brings to the electrical power system by virtue of a fully realized analytics strategy. The unprecedented access to the immense and growing volumes of data now available to describe the electrical system and its consumers can provide powerful and nearly instantaneous insight. This insight not only improves the ability to optimize day-to-day operations, but in times of stress, is the core enabler of effective decision-making and critical communications. When faced with uncertain conditions of extreme weather, terrorism, or other disasters, the safety and continuity of reliable energy delivery is without measure.

One could hardly say that the grid is broken today, but system reliability and efficiency has degraded over the past several decades. And change has been slow to come. It is imperative that, as a society, we find ways to make the grid more resilient, secure, efficient, reliable, and capable of integrating with the lives of consumers. The technical innovations inherent in big data analytics for the smart grid are the first step and the future step.

– Carol L. Stimmel, Nederland, Colorado

About the Author

Carol L. Stimmel began working with "big data analytics" in 1991 while hacking code and modeling 3D systems for meteorological research—years before that combination of words ever became buzzword compliant. In those 23 years, she has spent the last 7 focusing on the energy industry, including smart grid data analytics, microgrids, home automation, data security and privacy, smart grid standards, and renewables generation. She has participated in emerging technology markets for the majority of her career, including engineering, designing new products, and providing market intelligence and analysis to utilities and other energy industry stakeholders.

Carol has owned and operated a digital forensics company, worked with cutting-edge entrepreneurial teams; co-authored a standard text on organizational management, *The Manager Pool*; and held leadership roles with Gartner, E Source, Tendril, and Navigant Research. She is the founder and CEO of the research and consulting sustainability company, Manifest Mind, LLC, which brings rigorous, action-based insight to advanced technology projects that create and maintain healthy ecosystems for people and the environment. Carol holds a BA in Philosophy from Randolph-Macon Woman's College.

Acknowledgments

Most people will only glance at this section of the book long enough to notice that I forgot to acknowledge them, and they will likely be somewhat irritated or hurt. So, for everyone who indulged my absence from important events, forgave me for ignoring their messages, and accepted my mission to work on this book, thank you for your gracious understanding. To my loyal friends who consistently offered encouragement and comfort, I hope you have some vague idea of how much I appreciate you. And, to those who were impatient or lectured me about working too hard, perhaps you kept me humble, but I am still ignoring you. But mostly, I would like to acknowledge my dear friend Argot, who read every word of this book, gave up her nights and weekends to help me, made me laugh, and challenged me to be better—my enduring and heartfelt gratitude.

And with that, I will call it good.

Section One

The Transformative Power of Data Analytics

Chapter One

Putting the Smarts
in the Smart Grid

Analog computing machine in fuel systems building. (*Source:* NASA[1])

1.1 Chapter Goal

Smart grid data analytics are playing an increasingly critical role in the business and physical operations of delivering electricity and managing consumption. And even though utilities are starting from a difficult position with integrating data analytics into the enterprise, data science is a critical function if the

[1] Image retrieved from the public domain at http://commons.wikimedia.org/wiki/
File:Analog_Computing_Machine_GPN-2000-000354.jpg.

3

mission of the modernized grid is to be achieved. This chapter describes the overall drivers for the smart grid; the key role of data analytics; the challenges of implementing those analytics; and why—without a comprehensive data analytics program—the expectation for a clean, reliable, and efficient grid will be impossible to achieve.

1.2 The Imperative for the Data-Driven Utility

When Hurricane Sandy tore through the United States' Atlantic and Northeast regions, it left as many as 8.5 million people across 21 states without power, in some cases for weeks. The challenging situation demonstrated the fragility of the electricity grid infrastructure, and the difficult restoration underscored an inescapable fact: The largest machine in the world is crumbling in a graceless display of accelerating decay. While a smart grid certainly cannot totally prevent outages during a natural disaster, its information infrastructure brings the promise of a new level of service to the customer during major disruptive events and to our daily lives. Yet, despite incremental improvements, the global electrical grid is plagued by a worsening trend of severe blackouts caused by the combined effect of aging infrastructure, high power demands, and natural events. In the US alone, the power system has experienced a major blackout about every 10 years since the 1960s, and power disruptions have increased steadily both in frequency and duration over the last decade.[2]

Compounding the situation, research and development spending began to stagnate in the US and Europe in the 1970s in response to oil price shocks. Investment largely turned toward identifying new fossil-fuel resources.[3] Significant investment only experienced an upturn when the American Recovery and Reinvestment Act of 2009 (ARRA) directed billions of dollars toward building a modernized electricity grid and subsidizing progressive technology deployments, renewable energy projects, and advanced battery systems. However, much of the damage had already been done. The lethal combination of a business-as-usual approach to grid management and a disregard for technology innovation left a diminished capability to rapidly meet reliability demands.

---

[2] Joe Eto, "Temporal Trends in U.S. Electricity Reliability" (October 2012), IEEE Smart Grid. Retrieved September 19, 2013, from http://smartgrid.ieee.org/october-2012/687-temporal-trends-in-u-s-electricity-reliability.

[3] Jan Martin Witte, "State and Trends of Public Energy and Electricity R&D: A Transatlantic Perspective" (2009), Global Public Policy Institute, Energy Policy Paper Series, Paper no. 4. Retrieved from http://www.gppi.net/fileadmin/gppi/GPPiPP4-Climate_RD_FINAL.pdf.

While other developed nations fare somewhat better than the US at keeping the lights on, especially those that rebuilt their infrastructure after the widespread devastation of the Second World War, decay and resiliency aren't the only important issues. Other pressing problems motivate the smart grid: There are more than a billion people in the world who do not have electricity, and as the impact of anthropogenic climate change grows more alarming, nations are coming together to mitigate the threat with greater conservation, efficiency, and renewable forms of generation. The smart grid enables all of these approaches to treat the issues of resource scarcity and power delivery around the world.

The relationship between electricity availability and economic health is inescapable. High-quality energy delivery service is an imperative for developed nations, especially given the high cost of outages, which has been estimated to reach billions of dollars in a single year. In energy poverty–stricken nations, billions rely on health-damaging, dirty, and polluting fuels, and they spend hours every day collecting fuel to meet basic needs, such as cooking—not lighting, heating, or cooling. There is an opportunity to fill this dire electrification gap with clean technologies that is economically, socially, and environmentally viable.

The strides that are made in smart power in the developed world can serve as a reference architecture for solutions across the globe. Despite the fact that demand will likely remain steady in developed nations due to increased building and vehicle efficiency improvements, demand in the developing nations is growing robustly as indicated by the US Energy Information Administration (EIA) (see Chart 1.1.), creating opportunities for intentional energy solutions that are sustainable.

Economic drivers, carbon reduction, regulatory compliance, and an increase in the drive to provide residential, commercial, and industrial customer self-management of energy costs and consumption are creating the perfect storm for grid modernization and smart electrification.

The current centralized model of power delivery, with its fragile, legacy, and manual componentry, simply cannot accommodate energy and efficiency demands in the way that an intelligent, distributed power system can. An information-based grid solution that enables autonomous operation, efficiency, reliability, and higher power quality is the best solution we have for securing reliable electricity service and the energy future of global citizens. Smart grid technologies provide universal—and clean—electrification; alleviate climate change by enabling a variety of efficiencies and renewable generation; and get us closer to a guarantee of affordable, safe, and reliable electricity. To fully realize this mandate, utilities have no other course but to transform themselves into data-driven businesses.

As sensors, intelligent devices, advanced equipment, and distributed systems are integrated into the grid, various forms of data will empty into the

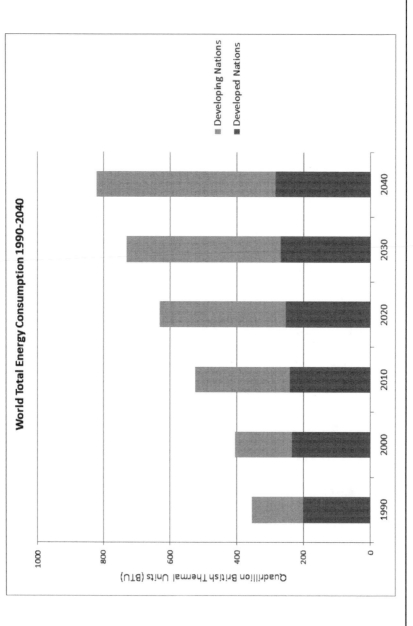

Chart 1.1 Energy Demand Forecast to Grow Robustly in Developing Nations. *Note:* BTU = British thermal unit. (*Source:* US Energy Information Administration, International Energy Outlook 2013)

utility at ever-increasing rates and volumes. With a carefully conceived, scalable approach, data analytics will quickly become the focal point to understanding real-time situations on the grid, past-event data, and the best and most efficient ways to meet customer needs, run a business, and improve both system design and performance.

1.3 Big Data: We'll Know It When We See It

The utility industry is just seeing the beginning of the growing number of bytes that will increasingly flow through the energy network—and it's unprecedented. Yet, defining "big data" is a constant cause of hand-wringing among those new to the concept, likely due to its buzzword status. It's a relative term and frustratingly imprecise; while it has been widely reported that worldwide we gather 2.5 quintillion bytes of information daily, that number is on track to grow by over 4,000 percent by 2020. And though many analysts like to "munge" meter end points and data packet sizes into estimates of data volume for the utility, that number—even if accurately calculated—isn't going to tell us much. "Big data" is just a way to describe a data question, a degree of difficulty, data management tools, data science problems, and the data sets themselves. While it was first described with the characteristics for which Doug Laney of Gartner Research is initially ascribed, it is now cleverly called "3V," with the three V's representing volume, velocity, and variety. Notably, there is an unofficial fourth V: value.

McKinsey and Company takes the narrative route to describing big data: "[It] refers to datasets whose size is beyond the ability of typical database software tools to capture, store, manage, and analyze." And further, "[T]he definition can vary by sector, depending on what kinds of software tools are commonly available and what size of datasets are common in a particular industry."[4] So, in essence, when an organization's data gets so voluminous that it starts to cause problems, then it becomes "big data." I prefer the words of former US Associate Supreme Court Justice Potter Stewart in an opinion made in an obscenity case, "I know it when I see it." Given the ever-increasing scope of data collection, a reflexive definition may indeed be the most useful: Big data is big data, and it's getting bigger.

For the utility, harnessing these volumes of data means looking beyond legacy information sources to smart meters, digital sensors and control devices, wholesale market data, weather data, and even social media. The breadth and

---

[4] James Manyika, Michael Chui, Brad Brown, and Jacques Bughin, "Big data: The next frontier for innovation, competition, and productivity" (May 2011), McKinsey and Company. Retrieved from http://www.citeulike.org/group/18242/article/9341321.

depth of data have been overwhelming to energy provider stakeholders. Yet there is progress. In the context of the smart grid, smart meter data has been the easiest to collect and manage, both technically and from a business perspective, because of it represents the lifeblood of the utility—meter to cash. Many vendors who first moved into the smart meter data management system (MDMS) space tend to have long-term utility relationships and are trusted providers for this very sensitive function, which has served to calm the data analytics jitters.

Although this should be a harbinger of good things to come, too many utilities with smart meters have not progressed to analyzing the more-granular meter data (which includes much more than consumption values) and have relied on monthly roll-ups of the smart meter data for easier integration with legacy systems, either spilling the leftovers on the floor or "saving it for later." Leading MDMS vendors are trying to push these laggards forward by nominally including analytic tools with their offerings. Buyer beware, though; these "analytics" often amount to nothing more than glorified reporting features, falling far short of the true promise of the power of data analytics to the utility and potentially undermining true data science efforts. Further, even when data analytics are usefully applied to smart meter data, the more-substantive value of these models only emerges when data from across the enterprise and third-party sources is fused together and leveraged in aggregate for maximum predictive strength.

1.4 What Are Data Analytics?

Similar to "big data," the term "analytics" is a neologism, bringing a new and confusing usage to a well-worn word. It pays to deeply understand what analytics are and how they're driven by data science. Compared to reports that are usually intended for the business stakeholder and answer very specific questions (in fact, report processes will often be tweaked repeatedly to precisely drive the kind of answer that is being sought), data analytics help raise and answer questions that have been unknown until the analysis is begun. In reality, analytic models could not be further from a spreadsheet and presentation layer that is often called the executive console.

> *Utility big data analytics are the application of techniques within the digital energy ecosystem that are designed to reveal insights that help explain, predict, and expose hidden opportunities to improve operational and business efficiency and to deliver real-world situational awareness.*

It's not as simple as picking up some data and churning out statistics. The analysis itself is just a piece of the whole smart grid data analytics puzzle. Before the daunting techniques such as data fusion, network analysis, cluster analysis, time-series analysis, and machine learning are even contemplated, the underlying data must be collected and organized. Collection itself is a challenge, given the wide variety of data available across the utility. Organizing data is where the coherence trial really begins. The process includes cleaning (fixing bad values, smoothing and filling in gaps), joining various data sets, and storing it all in a data warehouse of some type. Analysis can then begin, but even advanced analysis does not complete the picture. Once analyzed, the processed data must be presented to users in a functional and low-friction manner so that it improves actions and outcomes. Even squeaky-clean data and advanced analytical processes amount to nothing if the resulting information cannot be understood by the users, if conclusions can't be drawn, and if no action can be taken.

1.4.1 The Data Analytics Infrastructure

The advent of big data is putting stress on the familiar approaches to data handling. Extract, transform, and load (ETL) processes have been the bread and butter of data warehousing since the banking and telecommunications industry first adopted them. With ETL, data flows predictably from data source to data store in a controlled and reliable manner. Most simply, ETL is:

- *Extract.* Reading the data from a data source that could be in a variety of formats, including relational or raw data.
- *Transform.* Converting the extracted data from its current form into the form of the target database. Data is combined with other data according to a predetermined model.
- *Load.* Writing the data into the target data warehouse.

ETL is the gold standard when the handling of data needs to be consistent, repeatable, and tagged with a verifiable chain of custody. Traditionally, there are different systems for data generation, transformation, and consumption. However, for big data, the ETL infrastructure is expensive and doesn't scale as readily as new technologies—such as Hadoop, an open-source framework that allows for the distributed processing of large data sets across clusters of computers; and the "Swiss Army knife of the 21st century"—that support the ability to process, manage, and give users the ability to directly consume data without moving it around. New methods dramatically decrease data latency (no copying from system to system), additional hardware is not required, and software-licensing fees can be reduced. At the same time, ETL purists persuasively argue

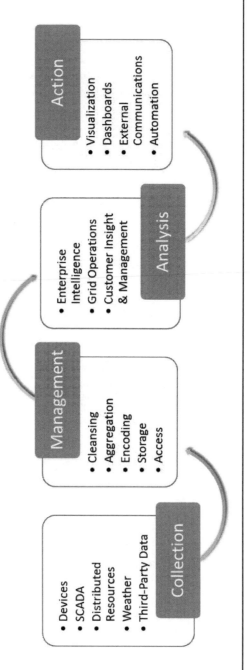

Figure 1.1 High-Level Flow of Data in the Utility.

that tools like Hadoop simply consolidate the steps within the ETL process to adapt to big data performance and scale requirements by running on a single engine. They further argue that the benefits of time-tested best practices should not be carelessly dismissed.

Figure 1.1 provides a high-level view of data movement from collection to action. The multimillion-dollar question to be answered for utilities implementing a comprehensive data management program to support advanced analytics is how that movement will materialize in both virtual and physical space.

1.5 Starting from Scratch

Despite the obvious advantages that data analytics bring to utility operations and customer service, most utilities are not using their data efficiently and effectively. The expense and monumental task of building an enterprisewide analytics program is intimidating, complex, and ongoing. Though utilities have made strides in sourcing and responding to data from smart meters, outage management systems, and supervisory control and data acquisition (SCADA) systems, only incremental progress has been made in using data and analytics to improve customer service, asset performance, network reliability, and operational efficiency.[5] When it comes to big data science, utilities are starting from a very weak position.

The temptation to introduce modest changes in the use of data analytics may be an imprudent choice. Grid modernization challenges not only the technical approach to power delivery but also the very foundations of a century-old business model. The current energy delivery business model is faltering with the new push for distributed energy resources (DERs) and new economic constructs that are being driven by technology changes; utilities that don't adapt to these shifts risk disintermediation, hollowing out, and ultimately an exodus of rate-paying customers.

Utilities that are moving forward now with implementing advanced technologies to support the evolving power delivery model will be best poised to meet the challenges of the significant addition of renewables to the generation mix, zero-emissions load balancing, and energy efficiency. Small data analytics pet projects and skunkworks are not enough to thrive in this new ecosystem. A step function forward in implementing smart grid data analytics is a requirement for continued reliable operation and business optimization in this new ecosystem.

[5] Oracle, "Utilities and Big Data: Accelerating the Drive to Value" (2013). Retrieved September 20, 2013, from http://www.oracle.com/us/dm/oracle-utilities-2013-1979214.pdf.

1.5.1 Mind the Gap

The significant expertise deficit related to big data management, analytics, and data science is one of the major reasons utilities have not been able to effectively use smart grid data. This problem is not unique to utilities, yet virtually every utility has this skills gap (in fact, almost every market sector that is data-driven is struggling). Specialized proficiencies are required to solve data problems, and up until very recently, very few academic programs have focused on big data and analytics. The pool of ready-to-go recruits just does not exist. With several million new jobs to be created in the field in the next several years, utilities, which historically aren't highly sought-after career destinations, are fighting for talent. Already, those who have recognized this deficit are scrambling to train current employees, recruit new people, outsource analytics to a third party, or invest in prepackaged analytics solutions.

But data science itself is not a simple discipline, which makes hiring to fill those roles in the highly specialized electricity industry, especially challenging. Data scientists not only need to know how to data wrangle, they must also know how to operate a variety of tools on a variety of platforms fed with vast amounts of varied data. On top of that, they must have business acumen and an understanding of arcane topics, like power engineering, energy markets, and demand response. Despite our best hopes, shrink-wrapped software is only going to go so far in solving the utility business problems. Energy-savvy data scientists are capable of changing the way the utility views the world and gets business done. Fundamentally, the point of data analysis is to carry us from raw data to information. And information is only available when the wheat is separated from the chaff and underlying patterns are exposed.

1.5.2 Culture Shift

Researchers have been raising the issue of information technology–operations technology convergence (IT-OT) for several years, and it's unhappily yet another term that defies industry standard definition. It's relative. Within the power industry, we often see stringent lines of demarcation between functions. The IT staff typically manages the transactional side of the enterprise: billing, accounting, asset management, human resources, and customer records. The OT side of the house manages the distribution operations, monitors infrastructure and control center–based systems, and oversees a lot of nonhuman interaction between systems on the grid. There just hasn't been an overwhelming need for chitchat between the system operators and corporate functionaries; this structure has met organizational objectives for decades, and the systems in place pattern themselves along these functional lines.

Now, grid modernization is driving not only technology changes but also business changes. IT and OT departments and systems must be integrated and work well together. There are many crosscutting business processes, and a lack of integration results in poor or uninformed decision-making, difficulty in meeting compliance requirements, poor communications, inefficient field operations, and the inability to effectively report to external stakeholders.

At the simplest, IT-OT convergence can mean allowing IT systems to dip into operational data, contributing to enterprise wide situational awareness. A key example is asset management analytics. Asset health models can be constructed by analyzing and seeking patterns in OT data based on information such as temperature, pressure, loading, and short-circuit, and fault events—all data that drives improved decision-making about how to manage a particular asset and conduct replacement scheduling. In fact, asset analytics characterizes one of the most important early smart grid wins for the business, reducing catastrophic outages while managing capital and maintenance outlays.

1.5.3 A Personal Case Study

Another, more personal illustration of how important IT-OT convergence is failing to handle major outage events. As I was writing this chapter, the lights went out. It wasn't unexpected, as I was in the middle of what had come to be called the 1,000-Year Flood in my home state of Colorado. As the monsoon rains came down, our electricity infrastructure failed us. Not only did this mean we didn't have lights, it also meant we didn't have information. Evacuations were occurring in our community, homes hung precariously from cliffs, people were trapped, and roads were collapsing. Losing power added to the intense fear, stress, and confusion we were all feeling.

As soon as the power cut, I loaded up the power company's website on my smartphone and reported the outage over the phone (the cellular connections were completely unaffected). At that time, I received an automatic message that the power was out in my vicinity and they were aware of the problem. The restoration time given to me was precisely 23 hours and 59 minutes from my call. Since the most direct route to our community was now a four-wheel-drive road and in danger of further rockslides, I was quite dubious. We spent the rest of the night with camping lanterns and flashlights and listened to the pounding rain—but happily with secure shelter. Nobody was complaining.

The next morning, the power was still out, so I rechecked the restoration time. There was now an additional 23 hours and 59 minutes tacked onto the originally posted time. I expected this trend to continue, as it was clear that the announced restoration times were arbitrary. Yet, later that day, three things happened:

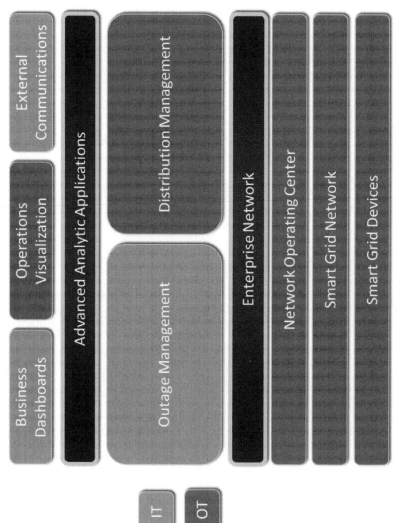

Figure 1.2 How IT-OT Can Converge to Address a Major Outage Event. Note: IT = information technology; OT = operational technology.

1. First, **the lights came back on** (easily inside of the initial 23-hour-and-59-minute window).
2. Two hours later, I received an automated phone call from the power company sharing "preliminary" information that **power would be restored shortly.**
3. An hour after that, another recorded call from the power company apologetically informed me that, because of the extreme damage from the storm, **it would be several days** before the roads would be passable and power restored.

This short story of the Colorado 1,000-Year Flood demonstrates how the failure to successfully converge utility systems results in poor situational comprehension and communication both internally and externally to the utility; in this case, blending the capabilities of the outage management system (OMS) on the IT flank and the DMS that resides with OT would have greatly improved the power company's crisis management.

Figure 1.2 describes a straw-man convergence of IT-OT systems. While OMS applications include business functions like crew management and trouble call management, the DMS applications perform the grid-facing operations such as fault isolation, switching, and state estimation. Both systems rest upon a shared network model that serves data to different applications across the enterprise. Grid-level data provides the best near-real-time situational intelligence across the enterprise, while IT systems make business sense of the information. As readily as the OT team can isolate the fault and estimate the recovery time, accurate communications can head out to news organizations, automated call routing, and social media.

1.5.4 Ouija Board Economics

In 2011, the Electric Power Research Institute (EPRI) estimated that an investment level of between USD $17 and $24 billion per year will be required for the next 20 years to bring the power delivery system to the performance levels of a fully realized smart grid. Using a complex cost allocation model, which included the infrastructure to integrate DERs and to achieve full customer connectivity, the model anticipated a benefit-to-cost ratio somewhere in the range of 2.8 to 6.0.[6] EPRI admits that the wide range in estimates underscores the uncertainty

[6] Electric Power Research Institute, "Estimating the Costs and Benefits of the Smart Grid" (2011). Retrieved September 24, 2013, from http://ipu.msu.edu/programs/MIGrid2011/presentations/pdfs/Reference Material - Estimating the Costs and Benefits of the Smart Grid.pdf.

in the industry in estimating expenditures and projected savings. The concept of the smart grid grows ever more expansive with new technologies, and the side effects created by the impact of generativity complicate the ability to get a hold on the problem—new issues emerge from making interconnections among old approaches—and the more innovation we bring to creating the smart grid, the less order and predictability there are in calculating an end state, whatever that is.

This lack of ability to prognosticate the costs to build a smarter grid has utility executives worried. Data management and analytics are sure to be one of the most challenging tasks for the utility, especially in scaling to the massive levels required to handle the sheer preponderance of anticipated data. This is apart from the related challenges of cybersecurity and data privacy. It is a difficult fact that despite the vital advantages that smart grid technologies bring to society, the required investments are massive for the utility to bear. In particular, generators will lose profits as a direct effect of smart grid–enabled demand-response initiatives, and economists do not yet have a grasp on how the benefits of smart grids can be easily converted into revenue.[7]

This may be a shock for utility leadership who have heard so much about the big data analytics opportunity, especially for improving operational metrics. While processes such as improving revenue protection and reducing asset maintenance and replacement costs are fairly straightforward, other functions such as assessments and improved planning are just not as clear-cut. With average utility investments in the smart grid rising to very significant levels over the next decade, many stakeholders expect a rapid return on investment (ROI), with more than half expecting a positive ROI in five years.[8] Taking current approaches, utilities are indeed engaging in wishful thinking, as they turn toward incrementally building up their capabilities to reduce risk and capital expenditures.

The desired level of return requires a more dramatic shift and efficiency of approaches to capitalize on smart grid data analytics opportunities in an industry that has a low level of experience around such programs. This is a tall order.

Put Your Head in the Cloud

Cloud computing and managed services are becoming a large component of big data initiatives, primarily as a strategy to help control costs and speed

---

[7] Luciano De Castro and Dutra Joisa, "Paying for the smart grid" (2013), *Energy Economics* (forthcoming). Retrieved September 24, 2013, from http://www.kellogg. northwestern.edu/faculty/decastro/htm/personal/payingsmartgrid.pdf.

[8] Oracle (2013).

up deployment time frames. Just a few short years ago, cloud computing was soundly rejected by the utility industry, largely for its perceived lack of security and inability to tailor software to particular needs. But customization is not viable for infrastructure providers because the economies of the cloud depend upon spreading costs out across the customer base as well as automating data and software management. Therefore, a service provider simply can't make bank if it focuses too much on custom development projects, rather than providing a menu of prebuilt options.

Some utilities are starting to understand these realities and leverage cloud computing with a more informed approach, although many still are not even considering these solutions, favoring an enterprise IT approach out of concern for security and control. This is mostly cultural and will begin to shift as utilities come to grips with the deep skills gap and the extraordinary capital outlay required to build up the required computing power and capacity to support comprehensive big data and analytics programs. Where computing scale is required, economies of scale are vital.

As a result, utilities must begin to look outside of their organizations into the cloud. Those that are seeking the benefits from cloud computing and managed services are likely to find them in improved speed of deployment, flexibility for meeting dynamic demand requirements, enhanced capacity, and most importantly, decreased capital expenditures. Most surprising to many utility stakeholders is that fact that cloud computing may actually offer a *more* secure and standards-compliant environment as service providers can deliver a harmonized approach with focused attention on cybersecurity and data privacy. Overall, cloud computing may be the key to giving utilities the opportunity to flexibly manage and deploy data analytics applications for rapidly growing data volumes in a secure and scalable manner.

1.5.5 Business as Usual Is Fatal to the Utility

Excessively conservative decision-making and low investment levels by utilities and regulators have created a slow pace of innovation for grid modernization. A bias toward proven and mature solutions has retarded the implementation of technologies that may ultimately be required for cost-effective operations. Regulators exacerbate this problem by excoriating utilities and denying them cost recovery when the uncertainties of novel investments are overtaken by risk sensitivity and political concern. This has long-term negative impacts. Without advanced systems and analytics controlling the network and subsequent improved decision-making, the elevated costs for managing the network will only go higher. As energy efficiency and distributed generation grow and consumption decreases, revenue will decline more quickly than delivery costs, resulting in revenue inadequacy. This alone sets the stage for spiraling rates and an exodus of customers who can buy energy from low-cost wholesalers, or self-generate.

Yet, the current approach to power delivery is not economically viable. Regulators and utilities must consider new cost-recovery approaches or risk disintermediation of the utility. The primary focus on curbing costs has driven down innovation, and that strategy has become very expensive. The risk of the hollowed-out utility underscores the importance of technology innovation and a higher risk profile when deploying smart grid data analytics. It is the smart grid infrastructure and the associated use of the data to inform better decisions that will ultimately decrease operational costs related to improved forecasting of demand, better ability for customers to manage their loads, enhanced service delivery and reliability, and an infrastructure that will allow new cost-recovery mechanisms.

1.5.6 To Be or Not to Be

Those with a more philosophical bent claim that utilities are facing an "existential crisis." That may sound like journalistic hyperbole, but utilities are indeed feeling the pressure of a bewildering collection of calls for enhanced energy services, including customer feedback tools, control and automation, cleaner energy, and customized rates and bundles for end-user applications such as rooftop solar and electric vehicles. They really don't know who they are anymore. While utilities begin to remake their business in various ways, they must fundamentally build a highly efficient and reliable infrastructure that can cheaply, reliably and efficiently deliver electrons. It is the advanced grid that will enable new opportunities, such as a move toward partnerships to deliver offerings through third parties or a complete restructuring to offer full-fledged energy services. No matter what ambition a utility has for its future, all emergent paths begin with a smarter grid and the enabling technologies that are found in advanced data analytics. Even a commodity approach to electricity delivery requires advanced systems that guarantee a platform upon which innovation can take place.

The following Figure 1.3 describes this continuum at a high level. It is most probable that energy providers will experiment and over time adopt a variety of these approaches. It is quite clear that to adapt the network to the two-way flow of energy and extend the network to meet the societal demands previously enumerated, energy providers must make a bold move toward infrastructure rearchitecture and alter how they make decisions across the enterprise—even

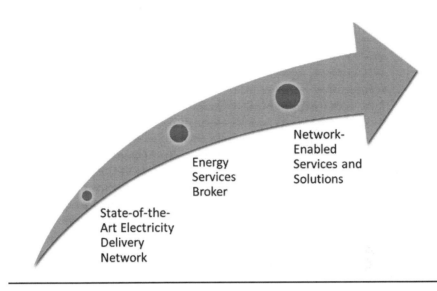

Network-
Enabled
Services and
Solutions

Energy
Services
Broker

State-of-the-
Art Electricity
Delivery
Network

Figure 1.3   The Evolving Utility Built on a Smarter Grid.

how they conduct even their most basic business operations. With the sensitivity toward risk management that is inherent in the industry, bold doesn't necessarily have to mean destabilization. At the same time, utilities should be wary of the call for incremental improvement. This is often just a way of hiding behind the fact that many in the industry have a superficial understanding of the issues.

This superficiality of understanding is an honest condition in the industry due to the politicization of energy; climate change, subsidies to energy companies, renewable energy, the business threat of microgeneration, the role of fossil fuels, and nuclear generation are all polarizing, bumper-sticker issues. With uncertainty surrounding whether these forces will impact the evolving utility business, it is no surprise that stakeholders favor an incremental approach. Unfortunately, this short-range vision ripples through to the private sector, making it very difficult to create affordable or innovative technology products or to ramp up production when utility buyers are so difficult to predict. Further, when private equity is willing to step in and invest in new technologies, a lack of foresight and extreme risk avoidance on the part of policymakers create lost opportunities.

1.6 Finding Opportunity with Smart Grid Data Analytics

Given the weak starting position of energy providers from the perspective of skills; cash; cultural challenges; and social, political, and regulatory forces, the rich and valuable flowing data will continue to pool on the floor until stakeholders find a place to start that provides an early, quick win but anticipates a comprehensive strategy. This requires business planning, a perceptual adjustment, and acceptance of innovation (and the special character of innovators). To realize the hoped-for societal gains, economic benefits, and ROI from the colossal investments being made in the smart grid, true data science must be applied to solving utility challenges both known and new.

Finding the low-hanging fruit is a compelling place to begin, as it helps build confidence in the role of data analytics with quick results and drives a fundamental understanding of big data, which is necessary for long-range strategic planning. Some utilities have found early value in analyzing consumption data to improve customer segmentation for improved demand-response targeting, revenue protection, and demand forecasting. A few system operators are already utilizing powerful data visualization tools to dramatically improve operational intelligence and management of their grid. And asset and finance managers are getting an improved handle on the impact of distributed generation on the system as well as the effects on revenue.

All of these early steps demonstrate the critical role that smart grid data analytics brings to bear on the business of delivering electricity. Data analytics and scientific innovation are changing attitudes and operations; fully realized, they are the foundation for the future of the electric power grid and clean, reliable, high-quality power across the globe.

Chapter Two

Building the Foundation for Data Analytics

One of the octagonal solar panels on the Phoenix Mars Lander. (*Source:* NASA[1])

2.1 Chapter Goal

The digital infrastructure of the smart grid is changing the nature of the power industry, and advanced analytics are the lever to realize the benefits of this intelligent structure. Traditional approaches to data management and to managing data truth in the enterprise fall short, where new software approaches can bring success. This chapter discusses the challenges of creating a highly scalable, easily managed, secure foundation for data management, and it explores techniques

[1] Image retrieved from the public domain at http://www.nasa.gov/mission_pages/phoenix/images/press/SS000RAD_CYL_P_10C70_R111T2_full_001.html.

that can drive the transformation necessary to meet society's energy demands with a modernized grid.

2.2 Perseverance Is the Most Important Tool

Marilyn Monroe said, "My work is the only ground I've ever had to stand on. To put it bluntly, I seem to have a whole superstructure with no foundation, but I'm working on the foundation." The utility that is working to implement a full range of smart grid technologies understands this problem. Utilities have been delivering reliable and safe electricity in a complex environment for decades. But now they're grappling with how to create an entirely new technology infrastructure on a sophisticated, yet insufficient, baseline that cannot have a moment of fault. Serious mistakes and missteps don't just upset customers who can't charge their smartphones; they have the potential to shut down critical infrastructure and cause severe economic disruption.

Making it even more difficult to solve this problem, vendors and innovators in the big-data science space are hoping to save the day (and make some serious coin), yet they do not always seem to understand the unique challenges of the utility industry. This creates strained relationships, slows the pace of implementation, and leaves utilities in the wild, scrambling to evaluate new technologies they may not fully understand, assess the consequences of their plans and decisions, and identify whole-life costs. In some cases, this slow pace has even drummed out hopeful start-ups that couldn't support such a long deployment cycle. This creates a difficult market environment. While partnerships are crucial in bringing the modernized grid to fruition through smart grid data analytics, the project requires the challenging step of building trust between old warhorses that can perform a Fourier transform in their sleep and a class of agile entrepreneurs who are fast-moving and masters of the sound bite. As if the technology problems weren't hard enough, this cultural and social chasm is a major problem for the industry.

Building a sustainable foundation upon which to realize the benefits of the smart grid requires bringing the best minds from the utility, big data management, and data science worlds together. It's really up to utility leaders to create these vital relationships by helping their partners understand their mission and values. Despite remarkable success with numerous smart meter implementations, many utilities are trying to move forward with an overall smart grid strategy, but they're slowed by a seemingly perpetual science experiment of small, multiyear pilots, lab trials, and troubled implementations that stress and even destroy partner relationships.

Times are changing, and this means utilities can no longer go it alone, no matter how passionate and committed they may be. The talent needed to succeed will come from many players, including power engineers, business sponsors, project managers, implementers, entrepreneurs, traditional and nontraditional vendors, and their investors. Under the leadership of the utility, partners must rise above the concept of a "negotiated contribution" and work with these partners to develop a shared vision of what can be accomplished by combining talents under the auspices of a common mission.

As the world's most famous starlet understood, when pretense is put aside, it's never the wrong time to push forward to improve the fundamentals.

2.2.1 "It's Too Hard" Is Not an Answer

Clearly, "doing analytics" can be extremely complicated and costly; the barriers are both real and perceived. Fear is a completely reasonable response when faced with the challenge of achieving returns from high-performance analytics on the needed scale for a satisfactory return on investment (ROI) in a timely manner. Data science is hard, and so are governance, compliance, and securely managing sensitive data. There is no way around the fact that implementing analytics requires investment and commitment from the organization, with the only guarantee that there are no easy answers.

Apart from a strong character, the utility stakeholder who wants to drive the benefits of data analytics into the utility needs to:

- Implement infrastructure improvements
- Deploy and develop analytical software and models
- Hire data scientists who understand the domain
- Generate results

Most importantly, the utility must find a way to minimize risk so that an analytic program can be engaged that will reduce the scope of uncertainty in embarking on what for many is an entirely new undertaking. Analytics surely can enhance customer service, business performance, operations, and overall profitability, but long-term pain is likely if the underlying logical architecture is not designed from the outset to be flexible and scalable.

2.3 Building the Analytical Architecture

A fully transformed grid requires a digital infrastructure that won't drain the enterprise by overly focusing on information technology at the cost of analytically driven strategic initiatives. As such, a well-conceived analytical architecture is derived from both the business strategy and the characteristics of the network.

Before an appropriate data management solution can be comprehended, existing and future business needs must be identified, along with a complete assessment of the types of utility data that will be managed. Establishing this groundwork is essential to defining an architecture that can accommodate business requirements, measurement strategies, and the overall grid structure.

In the early phases of smart grid data analytics, utilities may be tempted to inventory existing data from various systems and design discrete questions about the business or operation that can be answered by the data. However, this approach fails to capitalize on one of the most important goals of data science: to uncover the unknown. It also encourages a philosophy of incremental improvement. This may be a reasonable approach due to budgeting cycles and project management interests, but it deserves pause and careful consideration. Incremental, or serial, approaches that are blind to the preponderance of data flowing into the utility will result in a costly patchwork of implementations that will not meet the needs of the utility over time.

Think about it: You'd never attempt to build a house without a set of plans that documents accurately and unambiguously what is to be built. Further, an expectant homeowner isn't likely to plan and build a kitchen and then attempt to construct a home around it, room by room. Worse, you're not going to ask your son to go away and design the bathroom, your wife to design the hall, and your daughter to lay out the dining room in the hopes all the rooms fit together in the end. The builder needs to understand all the defining features of the site, the structure, and its mechanisms. Only with that knowledge will an architectural blueprint convey adequate information to realize the design. These same principles should hold true when designing the analytical architecture.

As described in Figure 2.1, the sources of data and how they will be collected, stored, and organized defines how effectively that data may be analyzed and shared as information for both operations and enterprise functions. And it is clear that to be effective, any solution must have profound ability to scale with controlled costs: As utility data generation becomes cheaper with commoditized digital equipment, and as throughput increases with bigger pipes, only data management is the greatest constraint. A well-designed system helps diminish limitations and allows the utility to focus on the practice of data analysis, where a lack of appropriate action can be critical and costly.

Figure 2.1   Data Management Process.

2.3.1 The Art of Data Management

Data management is an extremely rich and complex topic in its own right and encompasses a number of professions and technical competencies. According to the professional data organization Data Management International (DAMA, www.dama.org), the full data management life cycle includes "the development and execution of architectures, policies, practices and procedures that properly manage the full data lifecycle needs of an enterprise." Data governance, architecture, security, quality, and deep technical data management issues, including the entire data management framework, are topically outside the scope of this book. However, certain data management topics lend a more complete understanding to the subject of data analytics as derived knowledge and information, and these are addressed in the following discussion.

2.3.2 Managing Big Data Is a Big Problem

As we try to dodge the barrage of acronyms inherent in technology, it is clear that enterprise architecture is a bucket of obscure references. It's hard for anyone who is not a specialist to even begin to conceive of a data architecture project, let alone grasp arcane terminology and concepts. The despairing result is to leave it to the experts.

Hopefully, the experts that the utility works with will know plenty about data management and the special requirements of the analytics architecture. It is not the intention of this book to choose vendor solutions but, rather, to inform an educated discussion about what the best solutions might look like. Predicting or recommending the right data management techniques without understanding the unique nature of any particular utility is a fool's game, especially as software vendors and integrators rush to provide new solutions and remake legacy approaches. The big data/analytics realm has clearly been recognized as a robust market opportunity. Setting the direction for an analytics program requires an informed perspective; to make good choices in selecting technology and technology partners, utility stakeholders must understand a little bit about data management challenges, solutions, and the potential issues and limitations of these approaches.

2.3.3 The Truth Won't Set You Free

In the regulated world of the utility, where there is very little room for error, the quest for order and compliance is the dominant driver. And where there are so many functions within the enterprise, enabling interdependence through data integration has significant traction. It is logical to think that enabling the data-driven enterprise is dependent on the veracity of the data. The utility sector has always been pragmatic, and there is nothing more sensible than wanting a trustworthy view of the business. As you'll see, when it comes to data management, the truth is a matter of perspective.

The Single Source of Truth (SSOT) is an information systems theory that asserts that every data element in the enterprise should be stored exactly once, preventing the possibility of a duplicate value somewhere in the distant enterprise that is out of date or incorrect. When a particular piece of data is required, SSOT defines where that data lives and how to get it. This is the uptight sibling of the Single *Version* of the Truth (SVOT), where multiple copies of data exist because of recognized data silos but are "resolved" when truth is requested. SSOT and SVOT are often confused with single *source* of data (SSOD), where enterprises consolidate data to serve as the canonical source of fresh and accurate data. SSOT and SVOT are intended to free locked-up data wherever they reside in disparate sources across the enterprise for the greater good of data accuracy and consistency.

Many utilities implement a combination of a data warehouse and a master data management (MDM) system to standardize the truth. In this case, the data warehouse is usually considered to be the SSOT or SVOT. Data warehouses allow for the aggregation or congregation of data from multiple sources (including other databases) to provide a common repository of the data regardless of its source, thus, the single source. The MDM system governs the master data residing in the warehouse by brokering data from multiple sources, removing duplicates, and cleansing data to ensure consistency and control.

Although there is a ubiquitous passion in the industry to create this golden source of truth out of chaos (an example of which is described in Figure 2.2), master data may not be a silver bullet. It forces a normal perspective on data that may or may not be correct for the myriad users who access the records, and it can slow down loading and access to real-time data. For example, while both technically "customers," the customer who pays the bill is a financial buyer to the revenue manager, and the customer whose house doesn't have power is a technical end point to the field technician. A rigid version of the truth doesn't easily accommodate this variance; it takes a lot of money and time to build this kind of system, and it can be equally expensive and complex to maintain and customize when the truth needs a schema change.

Many data management experts wonder aloud if it's even possible—or wise—to attempt to create an SSOT in the enterprise. And unfortunately, this persistent desire for truth draws attention and resources away from solving business problems. Stephen Colbert satirically captures this propensity well when he

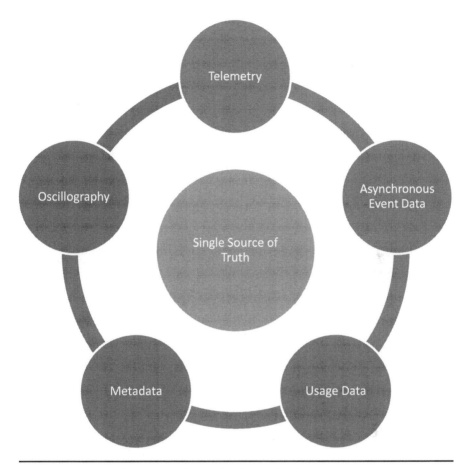

Figure 2.2   A Single Source of Truth for Smart Grid Devices and Systems.

defines "truthiness": "We're not talking about truth, we're talking about something that seems like truth—the truth we want to exist."[2]

As a way of managing data chaos, it may be less difficult and more advantageous to seek to provide context for the extremes in utility data and users. The real challenge for the utility is to become comfortable with the fact that the most orderly integration may still fail to correct all data integrity problems and will certainly fail to adequately serve all users. At the risk of sounding obvious, the utility's best choice is to consider architecting a system where the right data is delivered at the right time. This means eschewing orthodox data management

---

[2] "Stephen Colbert Has America by the Ballots" (n.d.), *New York Magazine*. Retrieved October 8, 2013, from http://nymag.com/news/politics/22322/index1.html.

practices for the benefit of advanced analytics problems where the data scientist is searching for something new and not yet understood or defined.

It can be uncomfortable to be dynamic and adaptive, but when core business practices are challenged, it is imperative. The art and science of data analytics is like listening to music: Why do audiophiles seem to appreciate vinyl so much over compressed digital audio? Vinyl offers higher fidelity, and the result is a more faithful rendition of its effect, technically because it contains the whole sound wave, which a digital signal can only approximate. If the utility standardizes and normalizes all enterprise data by pushing it into a single version of the truth (such as the data warehouse), the ability to find the remarkable signals in the data is lost. Data scientists are information anarchists of sorts, and they need the latitude to consider chaos and decide what is useful to them to help solve the problems at hand.

This is a sticky wicket for the utility: For transactional systems, governance, security, and predictable business process flow, unified versions of the truth may indeed be a worthy goal. But when it comes to advanced analytics, a single enterprise taxonomy will not suffice. Well-tested and predictable legacy data management approaches are in no way irrelevant; in fact, several integrators are finding ways to adapt existing systems to the needs of big data. The point is it may be time to shake off long-held assumptions about data management so that compliance does not inadvertently drive data irrelevance.

2.3.4 One Size Doesn't Fit All

Part of the reason this really matters in the domain of analytics is that one size will never fit all. Even in one organization. And this is terrifying to IT teams that are enjoined in a day-to-day struggle with technical debt. Stressed IT leaders will push the organization to build analytics on top of their existing stack and attempt to cobble together applications that will answer the business need for "analytics." Far short of true data science, this is an endemic problem in IT management, and many utilities will fall prey to standard solutions, such as:

1. Buy packaged software and customize it to fit the utility's needs.
2. Hire an integrator to build a custom solution.
3. Use cloud services with analytics applications as a way to "bridge the gap."
4. Do nothing.

Each of these approaches has its own concerns, including exorbitant deployment, maintenance, and operations costs; the "I built your software and now you're my hostage" situations; loss of control; or simply lost opportunity. These are not new problems, but analytics presents a special challenge; simply distributing data to users with standard processes will prevent an analytics program from realizing its full capabilities and benefits. In fact, it won't even come close. Analytics requires context-specific approaches to solve business problems as they arise, address operational issues in the moment, and find new efficiencies that can improve the bottom line and deliver a more reliable product.

It seems that we've put a lot of effort into figuring out what doesn't work effectively for a scalable analytics architecture that will serve the utility well over time. But, what may appear to be a fairly pessimistic view of application strategy actually helps lay the groundwork to confidently move forward with a course of action. Like people, technology evolves. The burgeoning analytics architecture in the utility is informed by all of the aforementioned responses, but we must shift our attention away from legacy or prepackaged solutions to discover how to deliver data to context-specific needs. This is really the only hope we have to meet the needs of high-performance analytics. And, yes, the beginning of that answer rests with another bunch of letters—API.

2.3.5 Solving the "Situation-Specific" Dilemma

API is the initialism for application programming interface (most people just say the letters A-P-I, as opposed to "app-ee," "ape-y," or "app-eye," which are a bit inelegant). API, once solely a technical term to describe how software components talk to each other, is now a way to describe a channel to retrieve the right data at the right time, wherever it resides and in whatever form it exists. It's really a system of software hooks that allow access to the data of interest residing in the underlying system, all without having to change the system infrastructure or a monolithic application. In essence, the API extends data to any authorized part of the organization (or to external parties) without ever exposing the underlying source code or master data. It is also the best mechanism for the utility to reach outside its confines to access third-party data to strengthen their analytic findings.

Integrating with the World

A story is begging to be told. This story describes how a company went from online retailer to an Infrastructure-as-a-Service (IaaS) powerhouse worth over USD $122 billion in 2013 by learning to effectively use APIs (sometimes called a "service interface"). This bit of private history became public when former Amazon engineer Steve Yegge posted an internal memo on Google+ with the wrong sharing setting. According to Yegge, one day in or around 2002, Jeff Bezos, CEO of Amazon, issued a so-called Big Mandate that was quite extensive, but included the following points:

- *All teams will henceforth expose their data and functionality through service interfaces.*
- *Teams must communicate with each other through these interfaces.*
- *There will be no other form of interprocess communication allowed ...*
- *Anyone who doesn't do this will be fired.*[3]

Those who did not refuse the gambit saw their stock options soar, as Amazon enables killer apps—including Reddit, Coursera, Flipboard, Fast Company, Foursquare, Netflix, Pinterest, and Airbnb—to be built over and over on its cloud infrastructure. No one except Jeff Bezos knows for sure why he had the foresight to push an online book retailer toward a service-oriented architecture (SOA), what is now simply called a "platform." Perhaps it was the realization that Amazon would never be all things to all people, or maybe it was just a grand experiment. Nevertheless, what became obvious was that a collection of various self-contained units of functionality collected into services could be combined in many ways to create multiple unique applications to meet the needs of the API consumer.

2.3.6 The Build-Versus-Buy War Rages On

No one is suggesting that utilities rush to deploy critical infrastructure or sensitive applications on the Amazon cloud. That's not the point. The point is that there is plenty of solid evidence that a platform approach—using a robust API—is a flexible and secure approach that can accommodate the fluid requirements of data analytics applications. In fact, it could—at the right price—be the ideal approach for analytics: flexible for the data consumer, easily secured, and able to accommodate both custom and packaged application projects.

The development approach to meeting unique application needs used to be called "bespoke software" (in reference to tailor-made clothing constructed to a user's specifications), and it allowed applications to be developed quickly, cheaply, and even more securely than commercial off-the-shelf (COTS) software or canned platform applications. Bespoke software gave us flexibility, but it also was often risky, expensive, and took a long time to develop. A better alternative

---

[3] John Furrier, "Google Engineer Accidently Shares His Internal Memo About Google + Platform," *SiliconANGLE*. Retrieved from http://siliconangle.com/furrier/2011/10/12/ google-engineer-accidently-shares-his-internal-memo-about-google-platform.

to both bespoke and COTS development is the concept of sharing components across many different business lines and across functional silos. Preconceived analytical applications will always require customization to meet utility needs, and it is an absolute that customization will be expensive, resource intensive, and likely to create unintended security loopholes. Figure 2.3 describes in general how sharing components changes the focus from the heavy lifting and security concerns of monolithic application development toward user needs.

Utilities have long used integrators for custom projects, and this often results in high initial costs for system design, coding, and deployment, as well as steep maintenance costs over the lifetime of the software; this is work that usually stays with the integrator because no one else really understands it. The expense and locked-in nature of custom development are at the root of the raging argument that has gone on for decades in business computing: build versus buy. Enterprises can't afford this level of custom development and maintenance, yet COTS is inflexible, expensive to customize, and hard to upgrade once customized. Neither choice provides the kind of rapid ROI required in today's financial ecosystem and crimps strategic innovation.

The demands of enterprise-class data management have the potential to lock out financially constrained utilities from the opportunity to benefit from an effective analytics strategy. This lack of ability to make the required expenditures can also slow the pace of innovation in the sector. Early big-data players are orienting their offerings toward more immediately lucrative opportunities such as the financial services industry, healthcare, government, and retail. The reason is obvious: The utility is slow moving and can be perceived as draconian. Furthermore, the diverse economic constructs and regulatory forces within the

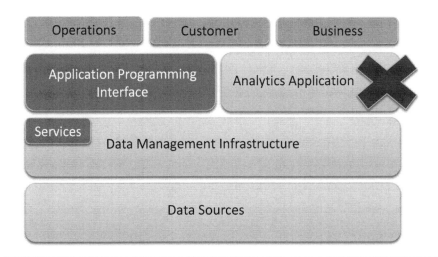

Figure 2.3 Example of Sharing Components Through a Service Layer.

industry are difficult to cope with; application providers and integrators have a hard time identifying the key industry requirements that allow them to build cost-effective solutions.

Given boundless financial, IT, and infrastructure resources, utilities would likely always select the bespoke way. But, a more reasonable solution from ROI, speed-to-integration, and long-term viability perspectives is packaged software that can be quickly customized to meet unique requirements (including financial constraints) without impacting a forward migration path. This is possible with the use of reusable application services that can be adapted and extended to create many distinctive applications.

Traditional bespoke projects require cadres of programmers who understand the underlying data structures, relationships, and workflows, but an application platform automatically handles arcane issues, including data quality, data consistency, and security. And it is by far more quality assured, consistent, and secure to solve these challenging issues in one place rather than within many applications. Programmers instead focus their skills on responding to business demands and building useful applications, including applications that can readily link to service layers from a variety of sources, converging and mashing data from disparate systems to create a richer, more valuable result.

Currently, there is one predominant model that meets these needs: a platform approach. A platform provides more than just packaged applications. A platform can be thought of as the foundation upon which to build applications complete with the required computing operations. This amounts to an analytics engine combined with a services layer, as well as a toolkit to integrate the custom applications built on those services. While a hosted infrastructure that provides facilities, power, and bandwidth is one way to achieve this, because of cultural and governance challenges, utilities are also experimenting with other methods to incorporate a platform approach, including licensing schemes and preloaded appliances that are managed for the utility but that physically reside within the enterprise.

No matter where it lives, this so-called Analytics-as-a-Service approach (some vendors use the AaaS acronym; we will avoid that, as we await a more clever marketing touch) is beginning to make a lot of sense. Some of these providers are even offering analytical packages that can be easily customized to help utilities get started with their analytics efforts, especially for more-normalized problems like advanced load profiling and meter analytics.

2.3.7 When the Cloud Makes Sense

These platform providers will encourage utilities to use cloud offerings because there are financial and operational benefits inherent in economies of scale for both parties. We have moved beyond the hype and the buzz, and the managed-services market has matured to the point where there is real opportunity to recognize business value from the cloud for both the provider and the customer. Noteworthy hosted offerings are working for many utilities that include tools to probe for greater demand-response opportunities by analyzing building-level profile data with publicly available consumer data, end-use disaggregation analytics, and very powerful visualization tools for grid operations. Utilities that first used cloud solutions to bridge the early gap with the intent to bring up their own in-house solutions are finding tremendous value, and over time the risks have been managed to meet their initial concerns.

Still, the reservations about handing over vital information to a cloud provider are neither insignificant nor unwarranted. While cloud solutions are gaining traction, there are reasonable issues related to losing direct control of sensitive data as well as meeting the requirement to comply with security and privacy mandates. Because they obscure visibility into how they secure and store data, managed-service providers are sometimes their own worst enemy. If these providers want to succeed, they must earn the trust of the utility sector by eliminating real vulnerabilities and ensuring that data is secured at every level. Cloud application and platform providers must be willing to undergo data security audits that prove their adherence to security standards. They should also perform regular penetration testing, maintain a track record of high availability, and provide disaster-tolerant data centers.

It is not surprising that utilities, as careful adopters of new technology, have been slow to embrace the cloud, preferring to manage the infrastructure and applications from within the enterprise. Even if the cloud is perfect for a start-up project looking for a leg up, when will it be a feasible option for the utility? It's complicated; the economic advantages have been slow to develop, but downward pressure is increasing and driving down costs. However, data governance, security, privacy, and loss of control over sensitive data continue to be barriers. The need for advanced analytics may be the turning point. With the imperative to drive smart grid data applications into the utility with a rapid ROI, the significant cost savings and nontrivial productivity boost make it impossible not to seriously consider the cloud as a viable option.

Figure 2.4 describes how an analytically oriented platform is approached, from source to the production of useful intelligence. In later chapters, we will dive deeper into the elements of the big data platform, analytics, and the transformation of data into actionable intelligence. For now, it is important to reflect upon the substantial operational competencies that are required at each level of the architecture. Only in a platform environment can the benefits of accumulated economies of scale exist.

The real issue for the utility is focus: Does the utility want to focus on expanding its information technology competences, or are there better long-

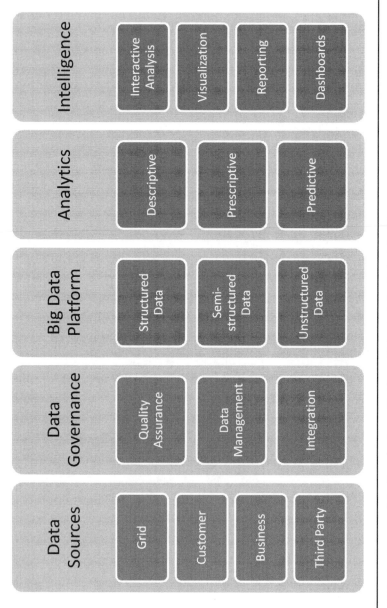

Figure 2.4 An Approach to Developing a Platform-Oriented Analytical Architecture.

term benefits to be achieved by focusing on business, reliable grid operations, and strategic issues? The answer to this question goes to the heart of the changing business model for the power delivery industry; utilities are likely to face a fork in the road where they must decide to be a wires-and-digital infrastructure for an intelligent network and begin developing and enabling partnerships that leverage that network, or turn toward bringing the capabilities online that will allow them to operate as a full-fledged service entity. A flexible approach to data analytics architecture will make the transition easier when that day arrives.

2.3.8 Change Is Danger and Opportunity

Building the foundation for an enduring data analytics program within the utility is not trivial. The entire scope of the company must be considered, including operations, business functions, and customer service operations. Unlike many IT projects, the determinations that utility stakeholders make about how to design their analytics architecture affect the future ability of the utility to do business in a cost-effective and efficient manner. Utilities in the era of the modernized grid are more than just poles-and-wires companies. Instead, they're at the center of a complicated and critical energy Internet (sometimes called the Enernet). Smart grid data analytics are changing the nature of how the utility makes every decision, and done poorly, it will inhibit the agility of the utility and prevent the realization of the effort to modernize the physical grid.

The science of smart grid data analytics is irrevocably altering the power industry, and it is simultaneously rife with possibility and risk. How load is balanced, outages detected and corrected, distributed energy resources integrated and managed, energy purchased—even the nature of energy demand—is changing with the need for greater reliability, financial constraints, automation, and the desire for more-accurate, but quicker, decisions. That is the superstructure that depends on a well-established data analytics foundation.

Some Questions to Ask When Thinking About a Data Analytics Strategy
How capable is our IT staff of taking on a major business initiative?
What kind of contractual commitments do we have for our existing data management solutions?
How are we prioritizing data analytics as they pertain to operations, business efficiencies, reliability, and flexibility?
How much does our current data management process cost us?
What will it cost to scale our process to meet the needs of our smart grid initiatives?

Chapter Three

Transforming Big Data for High-Value Action

Dr. Thomas O. Paine (center), NASA administrator, several NASA officials, and others applaud the successful splashdown of the Apollo 13 crewmen. (*Source:* NASA[1])

3.1 Chapter Goal

In this chapter, we build on the earlier concepts of the data analytics architecture and examine the role of algorithms and data presentment as well as the use

[1] Image retrieved from the public domain at http://images.jsc.nasa.gov/luceneweb/ fullimage.jsp?photoId=S70-35148.

of visualization to achieve a fully realized data analytics program. The goal of this chapter is to solidify for the reader the foundation for understanding how to use analytics to drive value from the smart grid. The topics introduced in this chapter are the keys for later discussions on applications and implementation issues. Smart grid data classes are also examined, as is the processing layer of the analytics architecture.

3.2 The Utility as a Data Company

Utilities are already skilled at acting upon familiar, structured data at a very large scale. Using data to make critical operational and business decisions is not new to the industry, and to assert otherwise is to deny the efforts that the utility makes every day in delivering reliable electricity. Managing resources, forecasting supply and demand, and administering demand-response programs, among many other functions, are all done today with the support of data analysis and automation. However, ongoing electrification goals, the significant addition of renewables, zero-emissions load balancing, flatlined revenue, and energy-efficiency mandates require more-accurate information, more-refined control, and tighter feedback. More-advanced forms of data analytics are required to achieve these goals.

While well-understood structured data will always be the foundation of a comprehensive data analytics program, much of the promise of smart grid data analytics based on the future of the more observable grid is the ability to leverage unfamiliar data sources and unstructured data. These forms of data, considered in aggregate, are very valuable for business analyses and operational rapid response. In many ways, the smart grid, with its variety of measurements, is an incredible opportunity for data scientists to make serious contributions to the industry and to society through key discoveries that improve operational and business advantages.

One of the most important goals of advanced analytics is to automate high-volume decisions across the utility, producing results that are dependably fast, accurate, and adaptable. As we discussed in the previous chapter, finding data analytics solutions based on the context and characteristics of a combination of many classes of data helps meet the needs of energy traders, transmission and distribution (T&D) operations, customer service, and the business. And with analytics, the sum is greater than the parts. As an example, transformer measurements have different uses depending on how they are analyzed: The data associated with the temperature of a substation transformer will be used differently in asset analytics management applications than the data from an operations alert that the transformer-monitoring nodes have identified anomalous measurements that require immediate action.

In the previous chapter, we discussed the challenges of developing an analytics architecture that includes flexible tools to help create flexible software for the enterprise. We also explored data management and integration approaches that help the data scientist make the most of every available data source. There are other important considerations that are required to ultimately transform data into actionable intelligence. These include the algorithms that drive improved predications and analyses and the transformation of those findings into actionable information.

3.2.1 Creating Results with the Pareto Principle

Vilfredo Pareto was an Italian economist and the father of the 80–20 (or Pareto) principle. First he observed that 80 percent of the land in Italy was owned by 20 percent of the population, and later he observed that 20 percent of the pea pods in his garden contained 80 percent of the peas. The 80–20 rule seems to have endless applications, such as 20 percent of the customers accounting for 80 percent of the sales, 80 percent of the results achieved by 20 percent of the group, and, of course, 20 percent of a company's staff accounting for 80 percent of the production.

For utilities that wish to quickly implement smart grid data analytics, we can see the Pareto principle operating in two distinct ways:

1. Algorithms designed to solve a smart grid problem will usually provide about 80 percent of what a particular utility needs, but they'll require 20 percent customization.
2. Data analytics applications designed specifically for the smart utility will be about 80 percent useful but will require 20 percent customization, especially when it comes to visualizing data.

The Pareto analysis has been used over and over again by management as a way to improve productivity, quality, and profitability by focusing on what matters most—mastering the 20 percent. For the utility, it reminds us to focus on finding and delivering quality and broad value in analytics as opposed to fixating on the infrastructure.

In fact, the Pareto principle teaches us little more than the fact that very few things in life are distributed evenly. It's an observation, not a law of nature, but it drives a focus on quality. For the utility that needs the most bang for the buck in a very short time frame, focusing in the critical 20 percent may be a business saver. Instead of trying to solve the entire analytics puzzle from scratch with the aim of unique perfection, utilities should create partnerships that allow them to focus on the strategic implementation of data analytics that help solve well-defined business problems. The Pareto principle doesn't mean only 80 percent of the work is valuable—all 100 percent is required in order for the program to work—but attention should only be paid to the efforts that will generate rapid results.

3.3 Algorithms

Using the broadest definition, an algorithm is a set of computational steps that takes input in the form of values and transforms those values into some output. It is the context of the problem that defines the relationship between the input and output. The algorithm itself describes the specific process for achieving that relationship. It is expected that the algorithm will produce the correct output (although incorrect algorithms do have some use, we will not concern ourselves with those here). The only real requirement for an algorithm is that it provides a precise enough description of the intended procedure to be carried out.[2] Obviously, as long as there have been problems, there have been algorithms to solve them. Computers just make it easier and faster, and they provide an almost infinite landscape of possible solutions.

Many of the smart grid–specific problems may be in their infancy, but as with any interesting problem, there are many candidate solutions based on what we already know about solving computational equations. Even the Human Genome Project, which has set out to identify all 100,000 genes that make up human DNA—a process that requires very sophisticated algorithms—uses well-known and well-understood methods to solve its problems. And it's not just data analysis that requires algorithmic treatment; the storage and management of massive data sets (3 billion chemical base pairs, in the case of our biological matter) also demand this step-based mathematical approach.

There are a huge number of classification schemes for algorithms. Classifying knowledge is never an easy task. Schemes for algorithm classification include purpose, complexity, design paradigm, and implementation. However, given the way utility data can be repurposed for many problems, architects will likely find it useful to classify algorithms according to the problem they're trying to solve. For example, an algorithm might use weather forecasting, seasonal variability, and demographic data as input data and create output in the form of optimal power-flow optimization metrics for managing a variable power source. This algorithmic approach could then be classified as a distributed generation process. That same input could also be used for customer demand modeling,

[2] Thomas H. Cormen, Charles E. Leiserson, Ronald L. Rivest, and Clifford Stein (2009), *Introduction to Algorithms, Third Edition*, Massachusetts Institute of Technology.

in which case it's then a demand management process. Such a classification approach helps break down functional silos in the grid to focus on business problems that solve operational, business, and customer problems.

3.3.1 The Business of Algorithms

In the business of data analytics for the smart grid, "algorithms" is also a term used to capture an aspect of a commercial product that's designed to address specific functional needs within the utility. Thus, when a utility is working to identify the best solution for its needs, understanding how a vendor implements algorithms is instructive in identifying the right tool or approach that will work across the system as well as the best application of data science to solve utility problems. Products may include "intelligent control algorithms" or "scheduling algorithms." The key is to try to understand algorithms from the perspective of how utilities use them, what their input and output are, and how their usage maps to a distinct business need.

One of the challenges with the purpose-based algorithmic viewpoint is properly framing the suite of utility problems. A business perspective is always the right place to begin and helps keep the focus on designing impactful solutions with measurable return on investment (ROI). As a starting point, classifying the characteristics of utility data available for problem solving is necessary to identify how that data can help solve the utility's business and operational needs. Taken holistically, this effort will drive an overall analytics scheme that defines not only what can be analyzed but also where and how to best collect that data.

3.3.2 Data Classes

A recent industry analysis of utility data showed the usefulness of grouping utility data into several classes and their characteristics. These five data classes are the basis of the data classes enumerated in Table 3.1.[3]

The business value of each of these data classes varies depending on how they're used across the utility, and as discussed previously, the underlying systems architecture will need to reflect those uses. Advanced data analytics often

---

[3] Jeffrey Taft, Paul De Martini, and Leonardo von Prellwitz (2012), *Utility Data Management & Intelligence: A Strategic Framework for Capturing Value from Data*, Cisco Systems, Inc. Retrieved October 4, 2013, http://www.cisco.com/web/strategy/docs/energy/managing_utility_data_intelligence.pdf.

Table 3.1 Five Classes of Utility Data

Data Type	Description	Functional Characteristics
Telemetry	Continuous flow measurements of grid equipment parameters and other grid variables	Telemetry allows the remote measurement and reporting of grid sensors. This kind of data is used for measurement that may be analyzed or used by control systems
Oscillographic	Data made up of voltage and current waveform samples that can create a graphical record	Oscillographic data may be pushed continuously or pulled through the communication network. The data is often consumed close to the collection point by other systems or may be carried back for postprocessing
Consumption data	This is most often smart meter data, but any node that measures usage data may be included	Consumption data is used for a variety of reasons, including billing and computing aspects of demand. This kind of data is collected and reported in varying time frames from seconds to days
Asynchronous event messages	Grid devices with embedded processors generating messages under a variety of conditions, both as responses and commands	By its very nature, this data is bursty. This class of data is challenging because the burst rate is undefined and many devices may respond to the same grid conditions
Metadata	Any data that is used to describe other data	Grid metadata is extremely varied and may include sensor information, location data, calibration data, node management data, and other device-unique information

Adapted from http://www.cisco.com/web/strategy/docs/energy/managing_utility_data_intelligence.pdf, with permission.

use data in new ways. Consumption data is a prime example of a data source that can be used to calculate real power (the actual power consumed by the load) to support operational requirements; it's also used for billing, to evaluate asset utilization and maintenance, and to inform overall planning. In fact, it is the ability to repurpose data that enables positive economic outcomes for the smart grid. Data that is used in as many ways as possible can support many outcomes and potential benefits.

3.3.3 *Just in Time*

Closely related to classes are the temporal aspects of data, or latency. Latency is the time delay of data movement within the system and is constrained by the maximum rate that information can be transmitted on a system as well as the amount of data that can be in motion at any particular time. Different operations have different tolerances to latency, and operations in particular may be sensitive to high latency. Every workflow is subject to latency, and, in fact, there may be multiple types of latency in effect during a particular system operation. Take a very simple example: It takes me 3.5 hours to travel by plane from my home in Colorado to Washington, DC. Even if there are 250 people on my plane, it's still going to take 3.5 hours. My latency does not change based on how many people are traveling with me. All 250 passengers will leave and arrive together. Once the plane lands and the crew is turning the jet around, the cleaning takes about 30 minutes and refueling takes about 15 minutes—altogether, 45 minutes of latency. However, if the cleaning and refueling happen at the same time, then my latency is reduced to 30 minutes, demonstrating, in some cases, latency can come at a bargain.

Latency considerations are extremely important in building a data analytics architecture. In fact, the failure to effectively manage latency can result in abject failure of the analytics program. If necessary data cannot be accessed in a timely manner to meet the goal of the analytics workflow, the application will fail. Because they can easily create a choke point in the system, data storage methodologies—coupled with latency—must be planned for and their dependencies avoided.

One of the biggest surprises to data scientists new to the utility system is the fact that many forms of grid data may have a data life span of microseconds, may never be logged, or may be overwritten at regular frequencies. Protection relay and senor data used in closed loops is used and then discarded. Also, telemetry data and asynchronous event messages may be stored in first-in, first-out (FIFO) queues or circular buffers where other applications may pick them up for use, stamping out the raw data as the queue or buffer refills. Transient data is very common, because only the freshest data is valuable in managing the state of the grid. In many utilities, only data that has been incorporated into business intelligence applications, or data that is subject to regulatory archival time frames, may be pulled into the data repositories or warehouses.

This is by design because no centralized data storage model will satisfy the needs of very low-latency controller systems. This data is dynamic, often used very close to the point of generation, and was never designed to be carried to the enterprise data center. Until the popularization of the smart grid and distributed generation processes, the utility system had been managed deterministically. Now stochastic models of operation are becoming the norm.

For analytics that involve operations, energy trading, real-time demand response, or asset management, analytic models assume that the necessary data will be available for some period of time. This is part of the data management challenge. The evolution of the grid from hierarchical to distributed, with the wide variety of data classes and latencies, creates an incredibly complex data-processing and analytics environment. Multiple schemes and a flexible architecture are required to accommodate the instantaneous triggering of actions based on rapid-fire analytics to long-range planning. These tools are being implemented now, and important new applications are taking advantage of this data across multiple business processes.

3.4 Seeing Intelligence

Leaving aside for the moment the vagaries of analytics designed for automation, the best way to transform big data and analytic opportunities and results into intelligence is a matter of context; what's most important is that the human being, who needs the information, can effectively understand, work with, and take appropriate action from the presented information. Transforming big data analytics into actionable information, especially with the complexities and demands for near-real-time awareness, requires the use of geospatial and visual modalities. In many cases, this is also true for downstream analytic applications as utilities work to comply with regulations, meet customer demands, and develop more-reliable services.

In the operational environment, access to distributed network sensors and assets has greatly simplified the detection of issues and situations on the grid. Though not widely deployed yet, grid operations systems that can process huge volumes of data in a range of formats and frequencies in real time are becoming a reality. Furthermore, by displaying disparate data classes in a visual and geospatial orientation, operators can see information across space and time to facilitate monitoring, rapid analysis, and action.

In the context of business intelligence, visualization is also an important tool, and nearly every vendor in the big data ecosystem that focuses on analytic tools and presentment provides some sort of visual access to data and intelligence. Within this emerging ecosystem, both horizontal and vertical platform providers are opening their end points to databases and end-user tool sets for maximum flexibility. In addition to visual environments, downstream smart grid data analytic applications enable reports, ad hoc queries, dashboards, other analytical models, and data exploration in a variety of presentment metaphors. Figure 3.1 describes how data flows from processing to presentment through a series of steps that helps best describe analyzed information.

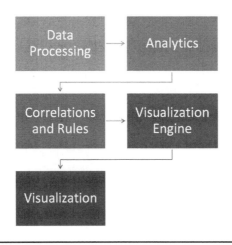

Figure 3.1 Process Flow from Data-Processing Layer to Presentment.

In the distributed environment and with heightened concern about efficiency and conservation, understanding relevant information is also important for the energy customer. Nano- and microgeneration are becoming more and more affordable for residential and commercial and industrial (C&I) customers. With C&I enterprises especially, these generation modes are growing in popularity because these customers cannot tolerate the financial exposure to low electric system quality or reliability. Additionally, demand-side management is quickly moving to customer devices, including computers, in-home automation devices, and the emergent Internet-of-Things (IoT) nodes. Even rudimentary technical solutions, such as the so-called Green Button initiative, that are designed to meet policy pushes to deliver energy data to consumers require relevant information. Early attempts to meet these customer needs are using analytics that combine utility consumption data with building-envelope data and home orchestration automation management and feedback information.

Automation is an important aspect of the visualization process. If the process for creating a particular visualization cannot be automated, then it simply cannot be scaled to many users or many delivery devices. Additionally, without automation, the systems must be continually updated and will likely lose their usefulness in a short period of time. Automation does bring risk, however, as source data changes or other bugs emerge. Just like the underlying data system, presentment systems must receive regular quality checks and attention must be paid to keep the information relevant and accurate.

It's essential to recognize that we are in the earliest stages of the evolution of big data in the utility. With the advent of smart grid data analytics in full enough force to economically enable widespread innovation, we will see data silos give way to intracompany data and data ecosystems. This shift is largely driven by the need for comprehensive situational intelligence.

3.4.1 Remember the Human Being

Some of the greatest challenges for effective presentment are found in the operations room. Unfortunately, there is a tendency for the industry to be entranced with information technology and advanced analytics, and to slough off the importance of the role that user-interface design plays in allowing a user to draw conclusions for quick and appropriate decision-making.

Designing intuitive systems that users can operate with a minimum of cognitive friction is the goal of user-interface designers who realize the stakes are high:

> *Managing the electricity grid is a complex job, and that complexity will only intensify as utilities incorporate two-way communicating meters, sensors, intelligent electronic devices, and myriad other technologies that monitor and report on the health of the grid. And while many utility stakeholders are worrying about establishing a return on investment from these modernization investments, engineers are working to make sense of a deluge of data that requires rapid response in changing situations. Without intuitive systems that provide good situational awareness, ineffective response (or the failure to respond at all) becomes more probable, and can lead to accidents with catastrophic consequences.[4]*

Incorporating the art and science of industrial design for user-interface design is critical and should be built into the project very early as a discrete part of the requirements process. User acceptance testing and the ability to quickly iterate on the design features will improve the quality and life span of the presentment portion of the application.

3.4.2 The Problem with Customers

Data visualization is hard. While the end result of effective data presentation is hopefully one that is simple and beautiful, often the process to get there is quite messy. The effort of building data visualization requires many skills beyond the analytic and statistical. It requires conceptual thinkers, graphic designers, programmers, user-interface designers, and good storytellers.

[4] Carol L. Stimmel (2012), "Smart Grid: Smart Grid Data Analytics in the Real World," *Smart Grid News*. Retrieved from http://www.smartgridnews.com/artman/publish/ Delivery_Asset_Management/Smart-grid-data-analytics-in-the-real-world-4967. html#.Um6dw2RilYg. Reprinted with permission.

In addition to telling a compelling and actionable story to grid operators and utility business stakeholders, utilities now need to present consumption data and other variables to consumers in the hopes that they will introduce more-efficient end uses and conservation strategies. The challenges in communicating data to consumers, especially, are a sore spot with some utility stakeholders. For all the effort put into building the smart grid and attempting to effectively respond to the sea change in the energy industry, it is hard to understand why things aren't working that well when it comes to consumers.

Here is some insight: In a recent online forum, a rather heated discussion about consumer engagement was taking place. A retired utility engineer wrote, "Utilities are essentially large project management/financial teams that are responsible for specifying, testing, buying, installing, and operating billions of dollars in assets to serve millions of people in actual real time, with minimal ROI. . . . Smart Grid is a wave of innovation, following 100 years of earlier waves."[5]

While likely not his intention, this retiree's comment may hold the key as to why utilities are missing the mark when it comes to consumer tools and applications, and it's not lack of effort, will, or intellectual prowess. It's a lack of domain knowledge for an area of pursuit that is so far outside of the normal demands of utility personnel that it becomes a case of not knowing what you don't know. And utilities really don't know that much about the people who use their product. They haven't had to know.

Utilities have been referring to customers as ratepayers and meter end-points for as long as they have been collecting an energy tariff. And residential consumers especially—despite extensive educational, marketing, and product design efforts—stubbornly refuse to engage with utilities in any meaningful way. Many innovators are attempting to crack the residential consumer nut, but others are focusing on the opportunities found within the C&I sectors. For good reason, C&I customers consume the preponderance of electricity, and there are often financial rewards for participating in conservation, demand-response, and efficiency efforts.

As described in Figure 3.2, in the United States, energy demand is projected to lift most in the C&I sectors, driven largely by the recovering economy. Due to increased efficiency in space heating, lighting, and other large appliances, the residential sector is not expected to begin growing again until at least 2029, creating an important opportunity for innovation that brings advanced analytically based demand-response opportunities to C&I customers.

[5] R. Hayes (2013), "Consumers Need Much More Than Education" *LinkedIn Smart Grid Executive Forum*. Retrieved from http://www.linkedin.com/groupItem?view=&gid=17 15027&item=277903596&type=member&commentID=5800677694588866560&trk =hb_ntf_COMMENTED_ON_GROUP_DISCUSSION_YOU_COMMENTED_ ON#commentID_5800677694588866560. Reprinted with permission.

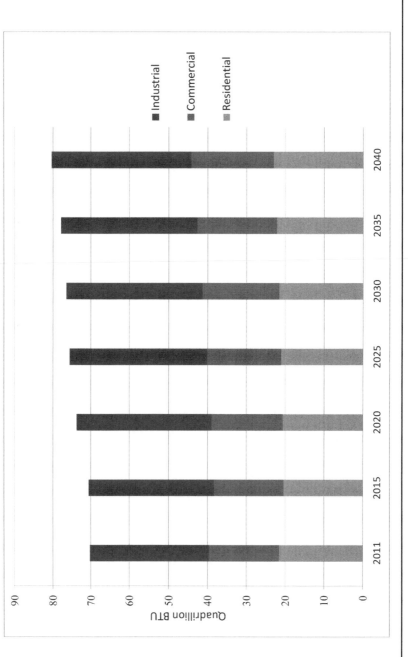

Figure 3.2 Commercial and Industrial Sectors Lead US Growth in Energy Demand. *Note:* BTU = British thermal unit. (*Source:* US Energy Information Administration, International Energy Outlook 2013)

Using advanced modeling tools, utilities can deliver actionable intelligence that allows demand management to transmute to engaged demand response, where end users can participate directly in utility dispatch strategies across a wide base of C&I facilities. The success of these systems ultimately depends solely on effective communication with these customers. Utilities have experience working with C&I customers on demand issues, and that expertise can be leveraged to drive further opportunity and innovation.

To capitalize on this opportunity, building owners, operators, and facility managers must be engaged to understand how they are using energy to learn how to participate in utility programs, analyze their buildings' energy use, and track consumption to realize energy efficiency and conservation goals. While analytics may provide the powerful measurements, correlation, and analytics necessary to advance demand response from an emergency tool to a strategic tool for managing business operations, an engaging and interactive presentation truly creates the opportunity for insight and action.

3.4.3 The Transformation of the Utility

Up to this point, we have discussed the drivers for the smart grid and how data analytics can help meet the goals of the modernized grid by transforming big data from the grid and other sources into transformative value for the utility. We have also reviewed traditional approaches to data management and discussed why these techniques may fall short in accommodating the full depth and breadth of the potential of data analytics. And we've identified the solutions inherent within a platform approach. But, at the end of the day, the real impact of analytics on the utility occurs when the applications enabled by the analytics architecture allow system users to see their business better and make more-informed choices. By answering key questions about the business and allowing exploration and interaction with presented information, data makes the journey to intelligence. Figure 3.3 describes how the analytics platform employs myriad processing strategies that are then presented to various users for an opportunity to explore the data to find answers to complex, multifaceted questions.

3.4.4 Bigger Is Not Always Better

One reminder: "Big data" is an industry fetish. Remember to think of big data as simply a description of a problem, meaning that there is much more data than the utility has coped with ever before. It is not the "bigness" that brings the kind of value that utilities need to justify the costs of building and operating the smart grid. In fact, the operational efficiencies gained by analytics are often

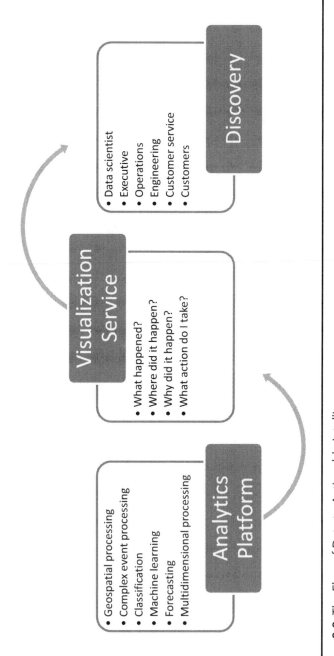

Figure 3.3 The Flow of Data to Actionable Intelligence.

from tiny and transient—not sizable—data sets. Big data flow, by nature of how it is collected, may be frequent and spurious, but the data itself may not be big, in and of itself.

In time, it will become clear that the petabytes and petabytes of data are simply a storage and processing problem with pure technology solutions in search of economic value. What's much more interesting are the challenges of meaningfully aggregating the disparate data sources that are produced at millisecond velocity to create a useful view of the grid for business, operations, and customer management. This gets back to the crucial point that the most important thing about designing and implementing data analytics is to identify the essential business questions. And then to acknowledge that all that big data doesn't actually make it any easier to formulate those questions—in fact, it makes it a lot harder.

3.5 Assessing the Business Issues

In the beginning, utilities are best served by assessing their overall analytic strategies and determining how those approaches will help them plan for growth and efficient operations. Understanding the data that the smart grid brings is undoubtedly the greatest issue that utilities are facing as they modernize their grid operations and business models. The information that's gained from analytics is the best opportunity that utilities have to improve business performance, energy quality and reliability, and customer relationships.

To begin this journey, utilities must look across the organization, above and beyond the current functional silos. The greatest value to emerge from data analytics is the aggregation of diverse sources of data from both the utility and third-party data sources. Additionally, utilities will not only discover how current business processes can be improved by bringing in other data sources but also prepare for emergent business cases. Although operational analytics can bring immediate value, thinking about the business holistically brings opportunities to experiment with new questions, such as adjusting and tweaking customer segmentation strategies by applying load profile data.

This is precisely why data analytics are transformative: They force utilities to reexamine all aspects of their business, from operations to customer engagement.

3.5.1 Start with a Framework

Determining the right solution based on business cases is a matter of breaking down the problem across functions. With a firm grasp on data management challenges and approaches, data classes, algorithms, and presentment, stakeholders can begin to map commercial capabilities to utility needs. There are several smart grid frameworks, reference architectures, and maturity models provided by the Institute of Electrical and Electronics Engineers (IEEE), National Institute of Standards and Technology (NIST), Carnegie Mellon University, and well-known integrators.

A very well-defined approach to developing analytics architectures, The Open Group Architecture Framework (TOGAF, www.opengroup.org/togaf) is used globally to define enterprise systems in government and Fortune 50 corporations. TOGAF has found exceptional success because it creates a common language for communication across the various skill sets required for the analytics practice. Specifically, TOGAF is maintained by a consensus and emphasizes key business imperatives that may be integrated into the technical architectures. In the utility market sector, vendors and utilities in the United States, Canada,

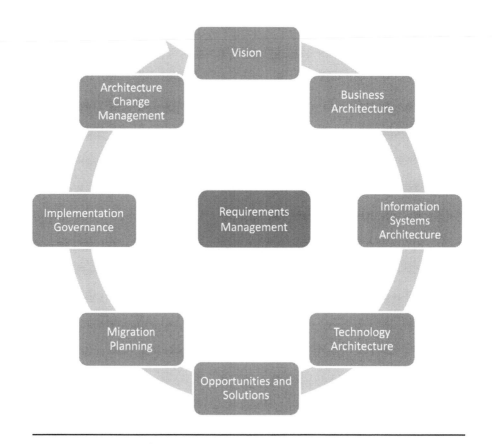

Figure 3.4 TOGAF Architecture Continuum Focuses on Requirements Management.

the United Kingdom, Australia, and other countries have implemented enterprise architectures using TOGAF.

The TOGAF methodology was originally derived from the US Department of Defense Technical Architecture Framework for Information Management (TAFIM), but has been adapted since 1995 to better serve enterprise architecture requirements. This methodology avoids proprietary methods, ensures consistent standards, and professes to help implementers realize a greater ROI with an open approach. Figure 3.4 describes the TOGAF continuum, with requirements management as an integral focus of each step along the way.

The characteristics of TOGAF are particularly suited to translate business needs and requirements into information technology requirements. However, as mentioned, there are several proprietary frameworks that are designed to specifically meet the needs of the smart grid–based utility enterprise. These frameworks (including TOGAF) are a way to guide thinking, but, as with any rigorous methodology, they can quickly become too complicated to use, or they themselves can become the focus of the effort rather than a tool to guide the outcome. Thus, these frameworks are best seen as organizational tools.

It's very difficult to start from a blank page, and architectural frameworks and reference architectures are a reasonable and informed starting point. However, be warned about an overzealous adherence to any framework, especially since a lack of knowledge about the desired end product—particularly in the case of analytics—can rope the organization into a flawed solution and can create a blind spot that prevents adaptation to new developments.

Widespread charges that utilities are dumb about their smart grid data are popular. However, while it may seem that it's just the data that is overwhelming the utility, it's even more challenging to figure out how to use the data to solve business problems. Data analysis just for the sake of doing analysis may actually bring negative results, with some stakeholders prematurely demanding expensive upgrades or redesigns to the overall business and operations. The way to manage a smart grid analytics program is to begin with a focus on answering high-value questions and on finding the necessary data to provide answers.

Framing Out the Problem
Define and document the most important key business measurements of performance in the organization.
If performance can be improved, what are the best ways to go about it?
How can data and information help us advance toward improving performance?
What is the set of goals that emerges to help define a data analytics business case?

Section Two

The Benefits of Smart Grid Data Analytics

Chapter Four

Applying Analytical Models in the Utility

The Apollo 1 crew in a parody portrait expressing their concerns about the fitness of their spacecraft. (*Source:* NASA[1])

4.1 Chapter Goal

This chapter introduces the analytical models specific to the utility enterprise, including the basic concept and goals of data modeling as well as the benefits and

[1] Image retrieved from the public domain at http://www.hq.nasa.gov/alsj/apollo1.jpg.

challenges of creating useful models. The use of appropriate analytical models is the foundation of process change within the utility to drive business value and return on smart grid investments. Additionally, we look at an optimization approach for taking steps toward balancing the forces of growth and profitability through the well-balanced and rational use of advanced analytical models.

4.2 Understanding Analytical Models

Uncertainties complicate our lives. As much as we might prefer a deterministic view of the world, we rarely have complete command of the consequences of everyday life. It's no different with analytic models, which must be flexible enough to provide strategic value under varying conditions. To make things simpler, we organize analytic approaches into categories. For the purposes of smart grid data analytics, we maintain four model categories: descriptive, diagnostic, predictive, and prescriptive. Many categorizations do not explicitly include diagnostic analytics, but given the operational requirements of the utility, it is important to specifically review this type of model and its role in the utility.

There are two important things about categorization: First, analytic systems rarely use only one category of analytics to produce useful results in a specific problem domain. Second, there is no real progression in terms of value from one category to the next. That means that despite the fact that predictive and prescriptive analytics may be more complex in design, they fill specific needs related to solving particular problems. Descriptive and diagnostic analytics may be better understood as a discipline, but they are not less valuable. Table 4.1 describes the analytic systems we will discuss in this chapter and how they function to help solve utility problems.

There are many examples of how these models can fit together to solve a business problem. Consider the case of an energy-efficiency program designer who is working to create a new offering that includes installing smart thermostats into homes for demand response. The utility is going to subsidize the rollout

Table 4.1 Analytic Models Used in Smart Grid Data Analytics

Analytic Approach	Function
Descriptive	What happened or what is happening now?
Diagnostic	Why did it happen or why is it happening now?
Predictive	What will happen next? What will happen under various conditions?
Prescriptive	What are the options to create the most optimal or high-value outcome?

with the ultimate goal of using automation to create a reliable source of demand-side relief for days of high stress and high consumption. This is an expensive endeavor for the utility; how can it determine which consumers will be interested in the program and which are likely to participate? Also, what are the best messages and incentives to encourage customers who may show a propensity for implementing conservation and efficiency measures in their homes?

Using analytic models, here are some steps a utility analyst could take to answer these questions:

1. **Descriptive modeling.** Of the customers who have previously participated in demand-response programs (such as a one-way pager switch installed on an air-conditioning unit), what happened? Did they answer surveys, cooperate with the setup of the equipment and its signals, did they override the response, how often did they override it? Tracking this information provides a basic understanding of customers who participate in demand-response programs.

2. **Diagnostic modeling.** Prudently, the analyst would then want to determine why certain customers behave in certain ways. Are they hardly ever home? What's the impact of the incentive on their overall bill? Did they sign up for the incentive but then resist providing the utility access to their equipment? Are they sensitive to temperature fluctuations? What was the weather like during the opt-out behavior? Did they express dislike for utility control mechanisms? At this point, the utility knows some characteristics about the customers who participate in the switch program, but they also have a sense of why they make some of the decisions they make in terms of their participation.

3. **Predictive modeling.** Having a sense of the what and why of consumer behavior from previously treated data, the analyst now has the appropriate inputs to devise a model that will attempt to predict how consumers will behave under certain conditions with smart thermostats placed in their homes. Specifically, under similar conditions, how can we expect consumers to respond to smart thermostats? Manipulating the variables in the model allows the analyst to create a precise segmentation of customers who are likely to embrace the utility's control of their home's smart thermostat. In fact, a comprehensive model can help identify segments of consumers that the analyst might never have considered before.

4. **Prescriptive modeling.** Finally, the analyst endeavors to understand what the next best steps are to take to drive program success. Based on what is now a deep understanding of customers who are likely to want to participate, a prescriptive model can provide insight into the best marketing or engagement strategies and their relative trade-offs for reaching the appropriate people.

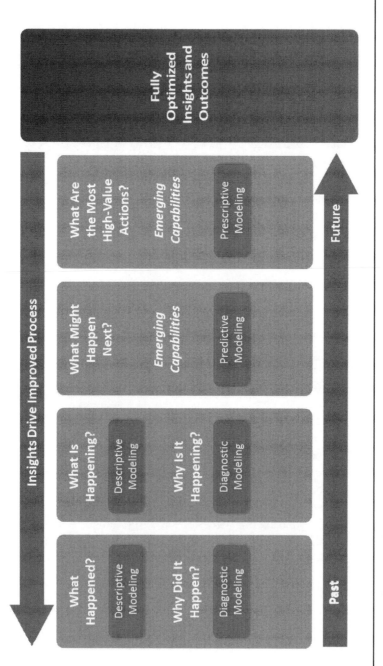

Figure 4.1 Analytic Models Use the Past to Prescribe the Best Business Actions.

Figure 4.1 describes how an analytics program can be structured to drive fully optimized business insights and outcomes. As insights are turned into action, these actions will change how the business operates (sometimes unexpectedly), and filter back through the analytical process in a cycle of continuous change and improvement. This is called generativity. The feedback loop of generativity may be one of the most important motivations for developing a comprehensive analytics program in the utility: *New structures and behaviors are already emerging under the forces of a shifting business paradigm; by analyzing these shifts, the utility can gain deep insight into new ways to improve the value and operation of the energy delivery network.*

4.2.1 What Exactly Are Models?

Models are the heart and lungs of advanced analytics. They use various algorithms and statistics to uncover the patterns and relationships we hope will bring increased value. But the best models aren't just the application of pure math. Yes, modeling is science, but it's an art, too. Like the master craftsman, the data scientist must have the ability to envision how the data pieces fit together; she must measure and construct a vision, and then produce something of long-lasting value that will be functional for the user.

To build a worthy model, the data scientist must be able to select the right data sources, algorithms, variables, and techniques that meet the needs of the business problem in question. These are the mechanical components, but they still require that the scientist have well-developed domain knowledge of the utilities enterprise. Both the development of the model and the communication of the model's results tell a story. The storyline is constructed by pulling through the right data to estimate and classify values.

Trust of the output that a model produces is perhaps the most difficult part of any analytics modeling process. For maximum perceived reliability, the model must reflect business realities—for example, showing how an asset maintenance model can drive down operating costs and demonstrating the value of the model's output. It is, in fact, a lack of business connection that can explain much of the fear and distrust of analytics. Sometimes this "value" is not always expected. A very good asset maintenance model may expose issues that the utility has not anticipated and has not grappled with, causing unforeseen workflow disruptions and expenditures.

It's worth pointing out that even a very good model is not some sort of mythical panacea; breakthrough discoveries are simply not a day-to-day expectation. Such expectations clearly defy the very definition of breakthrough. Models can be valuable for the organization even if they serve to reinforce and concretize implicit knowledge within the company. As discussed previously, this contributes greatly to improving and generating new insights for the system that makes up the utility business. To achieve this, the utility must hire or partner with data scientists who understand utility problems, data, and—perhaps even more importantly—utility processes and workflows that help them map analytical models to useful and trustworthy tools to improve the business.

4.2.2 Warning: Correlation Still Does Not Imply Causation

Part of the communication challenge for data modelers is helping analytic consumers understand the distinction between causation and correlation. These two terms are so often conflated that it can cause seemingly hopeless confusion. Causality and correlation confusion can disrupt the very goal of analytics, which is to transform correlation into causality. But sometimes there is a rush to provide an explanation for an observation, and that is done by claiming false causality.

By way of definition, correlation simply describes how two sets of data are related. Causation, on the other hand, defines a relationship between two sets of data wherein one creates the conditions for the other to occur. Consider the following commonplace example: A study shows that as ice cream sales increase, so does the rate of drowning, indicating that the consumption of ice cream causes drowning. We understand intuitively that this is foolish. However, this sort of leap happens frequently with less-understood data. In our example, we have not taken into account two important data points: time and temperature. Consider, then: Ice cream is sold at a higher rate during the warm months of summer than during the colder months. During these warmer months, more people engage in water-related activities, such as swimming and boating. The increased drowning rates are caused by an increase in human exposure to water during the same period of time that more ice cream is sold. This is the structure for a very common kind of causality fallacy called the "lurking variable"—a variable that, once known, disentangles the issue.

In fact, no matter how excellent a correlation may seem, there may be one of several logical fallacies that create false interpretations of data. When considering causation, it's helpful to think of the construct of cause and effect. Such as, when I throw a ball, it moves. Cause-and-effect relationships are rarely equally valid when inverted. Just because the ball moves, doesn't mean it was thrown. Think about it: There's a well-established causal relationship between obesity and an increased risk of gallstones. However, I may have gallstones and not be in the slightest bit overweight. The causative relationship is not true in either case.

No matter how profound the conclusion, it is obviously important to be wary of the logical pitfalls in rendering a series of correlations as causation. So the real question, it seems, is how do we safely conclude a causal relationship? It's difficult, but there are reliable methods we can use to prove that a correlation is causation, including randomized controlled experiments or the application of causal models. Put simply, the more powerful and robust we make the correlations, the easier it is to confidently draw a causative conclusion. Serious mistakes have happened, especially in medical science, where epidemiological studies have attempted to draw conclusions from data without fully understanding other factors that were responsible for the issue at hand.

4.3 Using Descriptive Models for Analytics

Using a descriptive model for analysis is somewhat like looking at life in the rearview mirror. We use descriptive analytics and techniques to understand what has happened and to also comprehend the deeper context of how something *may* have happened. We also deploy the descriptive model to support real-time analytic systems to understand what is happening in the moment.

In general, descriptive analytics explain source data in a way that allows the user to develop future business strategies. While "what happened" models are not normally used to model a precise event, they are useful in creating approximate perspectives from large quantities of data. It's not particularly useful to study why a single smart meter sent a last-gasp message, but it may be a very important data point if many meters of the same make and model consistently fail during read cycles.

In fact, descriptive smart meter analytics have already proved to be quite valuable for utilities that are searching for ways to use data to understand root causes with outside plant issues. In one case, after a major storm outage, the reporting system showed that there was extensive transformer damage. The utility wanted to understand more about the outages, and it determined with further data analysis that the root cause was breaking trees falling over the transformers during the storm. This information led the utility to make a strategic decision to identify at-risk transformers and provide treatment that would improve system performance and drive down corrective costs in future storm scenarios.[2]

Descriptive analytics are often relegated to the tombs of business intelligence. This is a gross understatement of the value of descriptive models. Imagine trying to understand a study about how to better engage utility customers where you're not presented with sample size, information about treatment groups, or

[2] Parmarth Naswa (June 12, 2013), "Analytics for Utilities," *Intelligent Utility*. Retrieved from http://www.intelligentutility.com/article/13/06/analytics-utilities.

demographic information such as sex or age. The descriptive model's output provides key summaries about the data and forms the basis of further quantitative analysis.

Descriptive analytics also form the basis for data summaries, which are a powerful way to understand a large set of observations. Consider a deployment of smart meters in an area that has a large number of service calls. To evaluate the performance of the system in this area, the smart meter outage statistics are collected and compiled, exposing low-voltage conditions that are creating brownouts and potentially damaging household electronics and appliances. This descriptive information becomes the basis of the decision to upgrade or reconfigure distribution lines within the area to improve reliability and quality.

Descriptive models are not well suited to expose the details of an event, and attempting to describe a large set of observations with a single indicator creates data distortion and the loss of important details. In a sense, descriptive analytics are self-limiting, but they do provide an essential summary of data that enables comparisons with data from other systems. Cross-referencing the output of a descriptive model that captures the characteristics of a collection of customer complaints by ZIP code with operational data is enough to provide better developed and more-varied models that support strategic decision-making within the utility.

4.4 Using Diagnostic Models for Analytics

Sometimes called inquisitive analytics, the use of diagnostic analytic models is closely coupled with descriptive ones. If you already asked the question about what happened or what's happening, your next question is why. Diagnostic analytics are subject to several of the same benefits and challenges as descriptive analytics, including concerns related to the issue of not having all the necessary data captured and available in order to come to the right conclusions.

Often, the questions of what and why are rather vague. For example, "Why is the customer backlash against smart meters gaining momentum?" The data that the utility may have available to answer this question could include structured information such as historical transactions with customers and billing data, but it could also include other sources of information such as news stories and social networking data, which are largely untapped. In fact, it is these outside sources that might be the real key to creating the full picture of customer dissatisfaction with the utility.

Though similar, the case for answering the question as to why something is happening is quite different in approach for descriptive analytics than for diagnostic. Rather than just assess the main features of the issue under consideration, a diagnostic model will further analyze the data to look for trends and patterns. Diagnostic analytic models actually attempt to use the available information to test and validate or reject hypotheses drawn from the descriptive analysis. Thus, the successful model will use drill-downs, factor analytics, and advanced statistical exploration.

4.4.1 How Diagnostic Tools Help Utilities

In North America, blackouts are most often caused by unexpected disturbances, most frequently on the wires. We count among these events hurricanes, floods, superstorms, and extreme heat conditions, creating issues such as control errors, coordination failures, and overloads. One of the most powerful tools we can use to gain insight into these disturbances is diagnostic analytics that help uncover activities that will improve reliability under these conditions.

Consider the Northeast blackout on August 14, 2003: The mega transmission failure was blamed on inadequate tree trimming, but that is only a piece of the story. As the tale goes, the grid operators did not have enough information to understand the extent of the problem at its inception and to mitigate it before it became a massive problem. With diagnostic tools in play to help provide situational awareness, corrective action could have taken place that would have avoided the surge-and-trip cascade. If there was any doubt before, the 2003 blackout demonstrated in earnest that excessive load on the transmission grid can lead to large-scale disaster. Knowing why there was lost voltage on the system (it was a hot summer day, the lines were overloaded, they were sagging close to the trees, and the breaker was tripping off the circuit) and being able to quickly understand and communicate that information would have allowed the grid operators to isolate the failure before it cascaded. New diagnostic tools, especially interactive visual tools, now facilitate oversight of intelligent routing with technologies such as syncrophasors, the ability to call on distributed resources, and the capability to trigger automated demand response to decrease grid stress.

4.5 Predictive Analytics

In ancient times, an oracle was believed to be a portal to the gods, a vehicle through which human beings could learn about the future, managing the insecurities of everyday life. These mediums could interpret the messages using tools such as bird behaviors and human entrails, prophesying in a frenzied state of passion during their consultations. Not surprisingly, the oracles held considerable influence in society, and their powers were widely sought out, and never doubted.

Very little seems to have changed in the world of prognostication. Predictive analytics is the use of advanced analytical models that are intended to answer the question—*What is likely to happen?*—to help people prepare for future possibilities. And it is partly true that the frenzy to leverage big data and analytic models to understand the future is just a modern-day manifestation of the impulse to put asunder human insight in favor of technical prognostication. At its worst, it is nothing more than a desire to believe that the world is governed by predictable events and averages. However, at its best, predictive analytics is a powerful tool to expose risks, uncover opportunities, and reveal relationships among myriad variables to guide better operational and business decision-making—certainly not a portal to the gods.

In the utility world, one high-value predictive analytics use case is the need for load balancing according to the day-to-day and hour-to-hour costs of power. The goal is to save both money and energy by predicting the costs of power and demand based on a constant flow of signals, allowing the distributors to buy and sell accordingly while shaving load during peak hours. The business problem is not new, but the approach that is enabled by predictive analytics is—specifically, creating an interaction layer between the bulk power system and the distribution systems.

An example of this kind of work was performed by the world's largest independent research and development organization, US-based Battelle. The Pacific Northwest Smart Grid Demonstration Project encompassed 11 utilities and tens of thousands of metered customers to create a system that engaged responsive assets throughout the power system to allow customers to voluntarily reduce energy use based on a control signal with data about power availability, price, and demand. The signal traveled throughout the system, altering the use and movement of power while driving down costs and simultaneously increasing opportunities to integrate reliability-challenging intermittent renewable sources of generation.

The project director, Ronald Melton PhD, explains how the system sent signals that communicated the actual cost of power delivery, to which the loads and energy resources could respond. For each communicating node, "a decision is made to increase the incentive signal value if less electric load is needed below that point, or decrease the incentive signal value if more electric load is needed. At the destination or end-use points, information about energy use is accumulated and forwarded to the source."[3]

---

[3] Ian B. Murphy (2012), "Utility Project Applies Predictive Analytics to Slice of Pacific Northwest Power Grid," *Data Informed*. Retrieved from http://data-informed.com/utility-project-applies-predictive-analytics-to-slice-of-pacific-northwest-power-grid.

This study demonstrates how predictive analytics can contribute to improving business outcomes by addressing a classic utility problem with an innovative technology-driven approach. By using a predictive analytics model with various inputs such as weather conditions, forecasts, fuel costs, historical usage, and other factors that impact renewable systems, an accurate prediction of production costs can drive the appropriate response throughout the system to meet both reliability and economic requirements.

Predictive analytics are best suited to well-understood and fairly stable situations, and they may perform poorly in situations where there is either little historical data or a significant possibility of rapid, dramatic change. Though predictive systems can be used to analyze data in flight or at rest, the utility will benefit from predictive systems that measure real-time information against historical data to identify fraud conditions, predict customer response to sales and marketing initiatives, forecast electricity demand to adjust production levels, and create a variety of risk profiles. Predictive data analytics improve with great volumes of high-quality data, and thus, the analytical model itself grows in value by evaluating disparate data sets such as weather; geographic information systems (GISs); and demographic, financial, sales, and social media data. Predictive applications for the utility include revenue protection, energy efficiency, program design, distributed generation integration and management (including revenue impact assessment), and demand-side management.

As mentioned, predictive analytics are powerful but not a silver bullet. Keep this in mind: When data analysts or vendors suggest that their model can predict an event accurately 75 percent of the time, they are still wrong a quarter of the time. Yet, 75 percent is far better than many of the models currently being used to solve very expensive problems. There is absolutely no excuse for an analyst, a vendor, or an enterprise to overestimate the abilities of predictive analytics, as such a position will bring disillusionment and a premature end to important efforts before they even get started. If your data experts are not communicating the level of predictive ability of their models to drive decisions, they are a liability to your organization. The most important thing to understand about prediction is the level of risk involved in trusting the model (trust is never without risk) and the consequences of action, if the model is wrong. Think about it: Seventy-five percent accuracy on a model that predicts the uptake of a demand-response initiative is a huge step forward, but it's a dangerous advisor when making threat-to-life or multibillion-dollar decisions.

4.6 Prescriptive Analytics

Prescriptive analytics are in their infancy, but they have significant promise. Taking the results of a predictive analysis, prescriptive analytics layer on a diagnostic model that ultimately produces recommendations for how to respond to likely events. In a sense, prescriptive analytics are the playing out of various predictions produced by changing variables to find the best decision in a particular context. The goal, simply, is to make a more informed guess about the most high-value action.

The merit of prescriptive analytics can be explained with a very simple example: I'm driving my very capable four-wheel-drive vehicle down an icy mountain canyon drive. I can see the road is in terrible shape, and my experience with this commute tells me the cars are driving too closely together and the conditions are ripe for a pileup. I slow down to increase the room between my vehicle and the other vehicles on the road, but just as I tap on the brakes, the car behind me slides forward and crashes into the rear of my vehicle. The force of the car hitting me causes me to slide, rolling me into the very cold creek. In this situation, my predictive powers were spot-on, and I did indeed successfully avoid an accident with the cars in front of me. Unfortunately, in the process of avoiding one accident, I caused another. Prescriptive analytics could have helped me make a better decision by analyzing the possible results of my actions and advising me of options that would not compromise my intentions: *Hey! You're being tailgated. If you put on your brakes, that fool may slide into you. Instead, turn on your flashers, and when you can, move slowly to the shoulder to let that guy behind you pass.*

Within the utility, it is anticipated that prescriptive analytics is especially valuable for taking preventive measures after outage-prone areas have been identified by earlier analytical models. Look to prescriptive analytics for answering the question of how best to do something—that's the key to optimization. For example, once we understand the context of our problem and its root cause, we need to know what to do: We know how many bills were reported in error last month due to meter malfunctions and where those units are located (descriptive); we have determined the root cause of the issue to be the failure of certain meters due to a spike in the readings and a posting of fatal error messages from a particular meter make and model that is installed in the field (diagnostic); and based on certain attributes of that make and model and what is installed in the field, we can predict when and where we are likely to see the next batch of failures (predictive). Additionally, with prescriptive analytics, we can create the most optimal plan to replace or repair those meters based on workforce constraints and loss of revenue, and we can plan to bring in further resources based on the financial consequences of not responding at the level we calculate we must.

The various utility models can solve almost any issue the utility is facing—if those models are applied correctly and with a reasonable understanding of the limitations of each of those approaches. Analytics are clearly more than throwing a bunch of tools at some aggregation of data. The right people with the right domain expertise are those who can use the tools correctly and are capable of communicating the results fairly and effectively, with a vital understanding of the abilities and liabilities inherent in their approach to solving the business problem at hand.

4.7 An Optimization Model for the Utility

"Optimization" is a crafty term used to describe the rather nebulous goal of achieving business perfection. Perhaps a more useful way to think about optimization is as an ongoing balancing act to meet the challenges of growth while maintaining profitability. In applied mathematics, optimization equations result in solutions where the controllable factors that determine the behavior of a system are minimized to avoid waste. In business terms, it's more like an evolving compromise between various conflicting requirements and demands. In the case of the utility, the smart grid is supposed to be the magic optimize! A red button that, when pushed, will defeat the extreme pressures to adapt more quickly to shifting political, business, and technology requirements while meeting increasing regulatory or downward cost pressures. In reality, utility optimization requires a process of discovery and the ability to respond to those discoveries by adopting new business processes that integrate continuous technological and business process advancements while controlling for risk.

Thus, while utility leaders repeatedly express a desire to make more and more decisions based on data analytics (in fact, most utility leaders state that they plan to increase investments in analytics technologies over the next several years), if analytic implementation strategies don't support business objectives, a lot of money stands to be wasted, and the result will be widespread disappointment and lackluster results. Instead, the introduction of analytic capabilities into the utility enterprise is best viewed as an enabler for change.

Utility optimization empowered by analytics is not simply a result of analytic tools and models to make better decisions; it's also the result of incremental advancements that occur from improving internal business processes based on what analytic results teach. Figure 4.2 describes the role that analytics play in creating optimal business value by striking a balance between growth and profitability. Once analytic capabilities are in play across the utility, process improvements will bring step changes in how the utility responds to operational events, business pressures, and regulatory shifts.

4.8 Toward Situational Intelligence

True situational intelligence is the foundation of an optimized utility, along the time continuum from hindsight to high-value future actions. Situational

Figure 4.2   Creating Business Value Through Analytics-Driven Process Change.

intelligence is the ability to derive understanding that allows the enterprise to make contextually relevant decisions with significant volumes of fast-moving and highly various forms of data. This means, as discussed previously, that data can and should be leveraged in many different ways across the organization in a manner that will help solve specific business problems. As described in Figure 4.3, the fundamental data intelligence layers that feed the utility are grid; meter; asset; and distributed energy resources, including renewables, microgrids, and storage.

Currently, data and organizational silos are preventing the emergence of these potential enhanced operating results across the organization, but real progress is being made in what can best be described as a bottoms-up approach to data analytics, where ad hoc teams attempt to piece together systems in the hopes of creating a more emergent system that will link many of these efforts. However, it is a top-down approach to analytics that will streamline the ability to solve business problems, instead of a web of expensive and fragile manual aggregation processes. Visualizing the entire network and the relationships within the network through a business lens will drive more accuracy in addressing real-time operational challenges, managing assets, and meeting performance metrics of every variety across the organization.

The smart grid is quickly becoming the de facto grid (let's just call it the smarter grid), not just in North America and Europe, but also around the

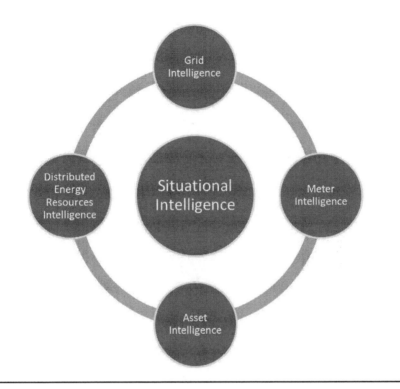

Figure 4.3   Origins of Utility Data That Promote Situational Intelligence.

world. Analytics are the key to unlocking the value of the smarter grid, communicating output from complex models to be useful and sharable, and ultimately driving the utility to be an information-enabled entity. With analytics, true situational intelligence emerges within applications that serve business functions, operations, customer management, and cybersecurity. We will discuss these applications in the next several chapters.

Chapter Five

Enterprise Analytics

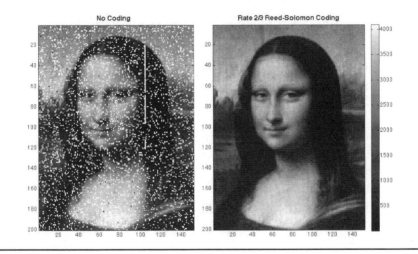

In 2013, NASA Goddard scientists transmitted an image of the *Mona Lisa* from Earth to the Lunar Reconnaissance Orbiter on the moon by piggybacking on laser pulses that routinely track the spacecraft. (*Source:* NASA[1])

5.1 Chapter Goal

In this chapter, we focus on the business-oriented, enterprise intelligence applications that are enabled by the various forms of data available within the utility. Specifically, we discuss traditional business functions that are enhanced by

[1] Image retrieved from the public domain at http://sciences.gsfc.nasa.gov/sed/images/featuredimage/featuredimage_317.png.

analytics, including energy forecasting, demand response, dynamic pricing, and revenue-protection analytics. These functions are separate from operations and customer management analytics, which are discussed in later chapters.

5.2 Moving Beyond Business Intelligence

Big data analytics for the enterprise include a new way of looking at what has traditionally been known as business intelligence (BI). BI is largely about generating standard reports that answer well-known questions, and though many BI vendors have attempted to expand the term to incorporate analytical capabilities, enterprise analytics are much more than a synonym for BI. Even though analytics have been used within BI applications, enterprise analytics are not only more sophisticated than a reporting system, they are more about the quantitative models that can be drawn from a deeper and broader availability of data, which can be analyzed and used to improve business performance. Analytics pioneer Thomas Davenport describes this potential succinctly: "The availability of all this data means that virtually every business or organizational activity can be viewed as a big-data problem or initiative."[2]

Many companies approach enterprise analytics from a simple access perspective, where employees across the enterprise have entry to available analytic tools for their own projects and where organizations can provision analytically derived reports and dashboards to employees. Other companies take an analytics approach, where multiple data sources, classes, and types are aggregated for use across the organization. With our focus on using analytics as an enabler to optimize the utility, the discussion of enterprise analytics occurs through the lens of those applications that are valuable and that are made available through the integration of disparate data sources. More specifically, we highlight the use of appropriate models to enhance traditional business processes, apart from operations and customer management.

The use of advanced analytics has a generative effect on the organization, and thus we are only scratching the surface of the tools and approaches that can be leveraged for utility optimization and new efficiencies. In fact, any analytical model that helps the utility meet customer demands while controlling costs is valuable. To help start the conversation, we discuss some of the high-impact applications, including energy forecasting, asset management, demand-side management, price modeling, and revenue protection.

---

[2] Thomas H. Davenport, International Institute for Analytics (September 13, 2012), "Enterprise Analytics: Optimize Performance, Process, and Decisions Through Big Data," FT Press Operations Management, Pearson Education, Kindle Edition, p. 4.

5.2.1 Energy Forecasting

Energy-forecasting analytic applications provide highly trusted and defensible load-forecasting models for short- to long-term planning horizons. These forecasts help the utility better plan and forecast resources, support energy-trading functions, and maximize return on investment (ROI) for smart meter and grid infrastructure.

Most simply, energy trading is the buying, selling, and moving of bulk electricity from the point of production to the point of use. As a commodity market, it is inherently volatile, and traders are motivated to operate as efficiently as possible. Moreover, the energy-trading business depends on an efficient end-to-end process, and undue complexity can lead to dramatically diminished results.

The drivers that impact energy trading and risk management are increasingly complex, especially as regulations call for additional trading of renewables. To meet these pressures, utilities require dependable, trusted analytical models that accurately predict demand with the ever-increasing influx of intermittent generation. And to support load-planning analyses that avoid expensive and inefficient trading miscalculations, intermittent generation necessitates granular forecasting capabilities.

Energy-forecasting analytics are really the foundation for addressing many of the business problems within the utility and can be best represented as optimization models that collect a variety of smart grid data operational sources, such as feeder demand profiles and capacity utilization data on a near-real-time basis. This data can be mined to support a multitude of advanced models that leverage the state of the network in a broad range of time horizons and scenarios. Especially in the enterprise, analytical models can be stacked and used in many contexts for many purposes. Energy-forecasting models benefit from a platform approach with a rich programming interface where common utility and third-party data sources can be accessed.

5.2.2 Asset Management

Asset management analytic applications—variously known as predictive asset management, preventive maintenance, or reliability-centered maintenance—help utilities run assets at peak performance and predict events that might cause unexpected and costly outages. Asset management analytics help reduce downtime, limit unscheduled maintenance, extend the useful life of assets, optimize maintenance cycles, and provide root-cause analysis for troubled assets. Advanced asset management systems may also provide automated monitoring and alerts as well as predictive capabilities for supporting asset maintenance and replacement decisions.

Because the utility is an asset-intensive organization, the value of such analytical power seems obvious. In the short term, though, asset management capabilities are a double-edged sword for utilities: There is a tremendous advantage to making asset-related decisions based on objective evidence; however, the costs of making unplanned expenditures based on a proactive strategy may be burdensome, as is the potential liability of exposing at-risk assets that ultimately fail because they were not tended to in a timely manner. But this is likely a short-term effect that will resolve as utilities increasingly rely on analytical systems to manage their resources and plan for these operations in financial and long-range models.

Asset maintenance is a strong example of what can be done in the utility to leverage a variety of data sources for a high-value ROI but also support multiple functional areas, including engineering, operations, business, and even field crews. Asset management applications combine grid sensor data with maintenance data; historical information; and specific information about any particular asset (such as inventory and warranty data) to perform asset monitoring, advanced model development, and root-cause analysis. The most immediate value of predictive asset maintenance analytics is the ability to detect failures and anomalies on equipment and to mitigate the problem before it causes an outage. Root-cause analysis is a key piece of this capability, since it enables engineers to perform targeted repairs and decrease time spent on troubleshooting in the field. Unfortunately, proactive maintenance is not always possible or feasible, but utilities that strive for an 80–20 (Pareto principle) ratio of proactive-to-reactive maintenance will fare well. Better analytical models are the foundational requirement for this achievement.

Gary Rackliffe, ABB vice president for smart grids in North America, explains the value of asset analytics in a conversation with the Utility Analytics Institute:

> There are a few dynamics that come into play in this new environment of improving asset health. First is regulatory compliance, which typically has very little gray area. Utilities need to inspect equipment at specific time intervals to comply with regulations. Additionally, there are two critical questions that utilities must address to maintain safe, reliable operations and to drive condition-based maintenance and asset investment decisions: what is the health of the asset, and how critical is the asset? These two questions enable utilities to determine asset total risk of failure. Criticality is an extremely important component. For example, the impact of an outage of a small cornfield distribution substation transformer is not as critical as an unplanned outage of a large generator step-up unit."[3]

---

[3] Mike Smith (July 31, 2013), "The Changing Face of Asset Management: Making Better Informed Decisions About Assets," *Utility Analytics Institute*. Retrieved from

Figure 5.1 shows the overview screen from the incumbent player in the enterprise analytics marketplace. The SAS Institute provides an asset reliability analytics tool as part of its suite of predictive asset maintenance solutions to help reduce the number of unplanned outages and to optimize repair and maintenance schedules. The dashboard view gives personnel the ability to visualize asset performance, with a focused view on a cost and capacity losses, detailed cost analysis, data drill-down by asset groups, and location-based visualization tools.

The use of asset analytics tools changes the process from a time-based maintenance practice to a data-driven, priority-based approach. But, as with the many challenges that analytics bring to the utility enterprise, with such a significant change, both processes and people will need to adapt to take full advantage of the benefits that asset analytics bring. Because the utility is particularly sensitive to risk, and because asset analytics clearly mitigate risk, moving analytics-driven asset management forward within the utility will gain earlier acceptance with strategic planners who are working on a long-term (sometimes decades-long) horizon rather than the tactical needs within a year-to-year window. Supporting analytic models that aggregate the available data will greatly improve this core function within the utility.

5.2.3 Demand Response and Energy Analytics

Managing peak demand is a challenge for nearly every utility, and, year over year, this problem gets more difficult to sustain the percentage reductions. Driving customer engagement with utility demand-response and energy-efficiency programs is a key part of meeting this challenge. Public utilities, especially in developed nations, have been trying for decades to successfully engage residential consumers with incentives, subsidies, and educational initiatives to encourage participation in these programs. Unfortunately, their messages are notoriously ineffective. Conversely, commercial and industrial (C&I) customers tend to be more tapped into these programs because they are granted significant financial incentives for participation; in fact, some companies make more money when they are paid by the utility to shut down or slow their operations than they do in the everyday functioning of their business. When grid reliability and quality are additive to the bottom line, meeting regulatory mandates and effectively implementing energy-efficiency and conservation measures are extremely important.

Theoretically, the aggregate of residential customers consumes a significant portion of the load, but it is the C&I sector that is more manageable and

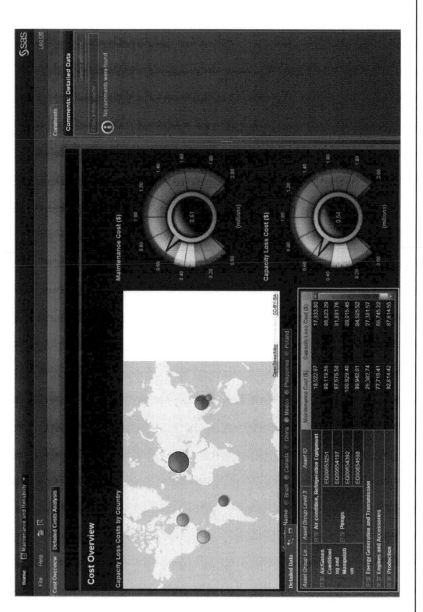

Figure 5.1 Reliability and Maintenance Cost-Overview Screen from SAS Asset Analytics. (Copyright SAS Institute Inc., Cary, NC, USA, All Rights Reserved. Used with permission.)

reliable. For example, shedding a single facility's load is equivalent to many homes, making only a single customer—instead of innumerable ratepayers— necessary to identify and reliably shed the necessary load. However, residential demand-side programs simply cannot be ignored, because these loads are increasingly representative of a disproportionate amount of consumption, and, historically, residential load drop can be delivered extremely quickly.[4] Analytics for demand-response applications may be the key to capturing the kind of reliable demand-side savings necessary to truly realize the potential for grid management, especially in periods of grid stress and high prices. Demand-response models help the utility plan and manage its programs. These applications enable utility program managers to identify both residential and C&I customers who are most likely to enroll, and, among those enrolled, who should be targeted and when.

Residential Demand-Response Analytics

It's costing billions of dollars worldwide to install smart meters, but the ROI, which includes the facilitation of improved consumption behaviors from customers' use of smart meters, is far from being realized. Early missteps in smart meter installations caused backlash among consumers and a heightened distrust of utilities, ranging from privacy matters to concerns about the health and safety of the meters. In the first rollouts, utilities were quite bullish on the beneficial impact that the devices would have for consumers; they were the jewel in the crown of the smart grid. After bills shot up for many consumers (due to several factors, including the fact that many of the replaced analog meters were running slowly), fires erupted, lawsuits were filed, and, finally, smart meter opt-out initiatives emerged that allowed consumers to keep their legacy meters. Utility business cases began to shy away from customer advantages and focused almost exclusively on the operational benefits for ROI.

Because the demand side of the conservation and efficiency equation is so important, this will not last. And as Warren Buffett is popularly credited with saying, "Only when the tide goes out do you discover who's been swimming naked." Utilities that leave out the consumer in their planning are without swimming trunks. Policymakers and regulators are beginning to demand that utilities explicitly consider the role that customers will play in the future

---

[4] R. Blake Young (August 28, 2013), "5 Reasons Why Residential Demand Response Matters," *Smart Grid News*. Retrieved from http://www.smartgridnews.com/artman/ publish/Technologies_Demand_Response/5-reasons-why-residential-demand-response-matters-5992.html#.Uo_PCmRiIYg.

of energy delivery. As a sign of the changing tides, in 2011 Canada's Ontario Energy Board published guidance and expectations regarding smart grid planning and investments and specifically called out the importance of customer control, education, and data access and underscored that plans would be evaluated and measured based on customer value.[5] Utilities that have avoided human contact must now find ways to effectively engage their consumers. Advanced analytical models that leverage both internal utility data and third-party data may be the most cost-effective way to succeed in an area that has historically proved to be very challenging for the utility enterprise.

That said, the problems of residential and small commercial engagement are substantively different than those of larger C&I entities. From an analytical perspective, C&I programs are focused on enabling responsive actions to price changes of energy over time, where it is financially beneficial for both the utility and the customer to interrupt operations during on-peak hours. Residential and smaller commercial enterprises—by virtue of the fact that they have less to give or do not have the latitude to shut down appliances like refrigerators, freezers, or home medical equipment—require a different approach. Some researchers and utilities believe that price-based mechanisms can be equally as effective with these customers, but in the short term, that assumption is questionable.

Many surveys have shown that even when customers are willing to participate in utility programs, they want electricity-bill savings that are considerably out of proportion to what can be expected in any sustained manner. In the US, this is especially ironic considering the excessive amount that utilities spend on customer incentives versus delivering actual bill savings to energy-efficiency and demand-response program participants. Developing new and innovative customer acquisition models is one of the key ways in which utilities can prioritize and begin to develop objective evidence about how best to engage customers, how to formulate a solid business case for utility programs, and how to estimate reasonable and reliable returns for their programs.

Disaggregation

One of the most promising energy-saving innovations designed to help bridge the gap between humans and technologies is disaggregation analytics. Disaggregation is the ability to use statistical approaches to treat either smart meter data or measurements from submetering sensors, exposing how

---

[5] Ontario Energy Board (February 11, 2013), "Report of the Board: Supplemental Report on Smart Grid," EB-2011-0004. Retrieved from http://www.ontarioenergyboard.ca/OEB/_Documents/EB-2011-0004/Supplemental_Report_on_Smart_Grid_20130211.pdf.

much power is being used by appliances, especially large devices such as air conditioners, water heaters, and furnaces. Clearly, this information can help customers understand how they are using their energy to identify savings opportunities and help them discover appliance inefficiencies and imminent equipment failures. Appliance-specific information is also valuable for research and development, improved utility-level sensitivity and accuracy in energy-efficiency and demand-response programs as well as load forecasting.

Disaggregation can be accomplished in two basic ways: monitoring a discrete load directly with an equipment-based solution using a device that measures spectral signature analysis or waveform-based analysis (both of which are highly accurate), or nonintrusive load monitoring (NILM), developed in the early 1980s at the Massachusetts Institute of Technology (MIT, US Patent 4,858,141), which uses analytics on the measurements of both reactive and real power to identify appliances by examining the voltage and current going into the house over time. Figure 5.2 shows a figure from the MIT patent published in 1989, showing how transient events can be detected and identifying discrete appliance start-up and shutdown events.

It's worth noting that NILM is far less accurate than hardware-based solutions, but it is much less costly and cumbersome, since it doesn't require customer intervention to be implemented with attendant management systems and support.[6] And while its modest results initially caused the technology to be largely ignored by the industry as a serious approach, with improved algorithms and analytical tools, it is now clear that NILM is adequate—maybe

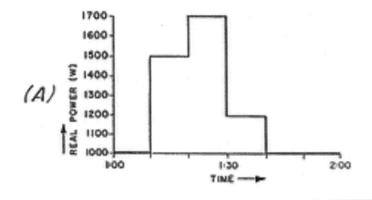

Figure 5.2 Model Transient Event Detection of Real Power (Watts) over Time. (*Source:* Figure 3 from US Patent 4,858,141, in the public domain)

---

[6] Jeff St. John (November 18, 2013), "Putting Energy Disaggregation Tech to the Test," *Greentech Media*. Retrieved from http://www.greentechmedia.com/articles/read/putting-energy-disaggregation-tech-to-the-test.

even a market-changer—for enabling the utility to maximize load planning, optimize energy pricing, improve demand-response and energy-efficiency programs, help integrate electric vehicles, and serve as a source of data for asset management applications.

Now that the value of effective disaggregation is being recognized within the industry, vendors are rushing to improve the algorithms that can be used in conjunction with smart meters, creating cost-effective and scalable ways of reaping the benefits of appliance-level usage data. Stanford University identified the various data features that are used by disaggregation algorithms to identify as many as 100 specific appliances. They include visually observable patterns, power transitions, and harmonics analysis to identity the type of electrical circuitry, transients, and appliance background noise. Disaggregation is surprisingly effective because, at different frequencies, various characteristics of appliance signatures can be identified.[7]

The potential of disaggregation analytics is close to full realization for smart meters. The cost to the customer is essentially nothing, the installation expenses and efforts are sunk, and the adoption rate will be virtually complete in the developed world in the next decade. Alternatively, hardware solutions with submeter capabilities for load monitoring are expensive, can be very difficult to install and manage, and have, so far, an abysmal adoption rate. Given this reality, the business case for analytic-driven NILM is compelling and potentially market changing when it comes to the myriad benefits that disaggregation can provide.

Commercial and Industrial Analytics

Serious demand response for the C&I sector has been the purview of aggregators, with the goal of automating participation. And it's big business. According to a report from the US-based PJM Interconnection (transmitting wholesale electricity in 13 eastern US states and the District of Columbia), USD $8.7 million in revenue was generated during just seven months in 2012 compared with $7.1 USD million in a 41-month period between 2008 and 2012.

As is often seen in cases of dramatic shifts in the industry, regulatory pressures drive change—in this case, if we dig at all, we'll find the Federal Energy Regulatory Commission's (FERC's) Order 745. This 2011 order mandates that, for generation and transmission, a demand-response resource must be paid the

---

[7] K. Carrie Armel, Abhay Gupta, Gireesh Shrimali, and Adrian Albert (2012), "Is Disaggregation the Holy Grail of Energy Efficiency? The Case of Electricity," *Precourt Energy Efficiency Center Technical Paper Series: PTP-2012-0501.* Retrieved from http://www.stanford.edu/group/peec/cgi-bin/docs/behavior/research/disaggregation-armel.pdf.

full wholesale price rather than the difference between the wholesale and the retail price. This changes the current approach, where the majority of the payments went to a small number of large customers that were able to commit more than 10 megawatts (MW) of shed.[8] How this order will ultimately serve to broaden the appeal of demand response has not yet been fully evaluated, but demand-response aggregators are working to shift their offerings. One thing is clear, though: Demand response in the large C&I market is enabling big business to participate directly in the energy and ancillary markets. And many of the tools—which provide customers with analysis and charting capabilities integrated with real-time and forecasted data from transmission organizations—are moving toward self-service.

One of the side effects of rapidly emerging innovation in the industry is the breaking down of barriers between utilities, grid operators, aggregators, and the end customers themselves. End-to-end management functionality is becoming the gold standard for utilities that are working to reduce or eliminate functional silos as part of their overall modernization program. Who will ultimately offer these services remains to be seen. While aggregators are angling to become full-service energy advisors that provide comprehensive software solutions, other utilities are working to bring this function back in-house.

Analytic capabilities both deeply embedded and as an enabling feature are key to moving from discrete demand-response applications to full-service offerings that tie together the customer and the utility as a single operating unit. Such approaches benefit both the utility and the customer by creating an operational system that hooks into customer management systems. With these tools, the utility can perform customer enrollment; manage programs; use analytical models for load-shed forecasting; optimize its portfolio; and even send notifications, automated signals, and postevent reporting to customers.

Clearly, predictive and prescriptive analytics that help the utility understand its high-value actions are the key value generator in an advanced demand-response system; yet, analytical models are working, albeit less obviously, in many ways, including generating forecasting and optimization models as well as user-friendly information for the C&I customer to understand its energy use and the benefits gained by participating in the utility program. Figure 5.3 is derived from a model described by the company AutoGrid for its Demand Response Optimization and Management System (DROMS) and explains the role that analytics plays in an end-to-end approach to demand response.[9]

---

[8] Katherine Tweed (April 2, 2013), "Demand Response Payments Increase Significantly in PJM," *Greentech Media*. Retrieved from http://www.greentechmedia.com/articles/read/demand-response-payments-up-significantly-in-pjm.

[9] AutoGrid, *AutoGrid DROMS*. Retrieved from http://www.auto-grid.com/technology/our-first-application-droms.

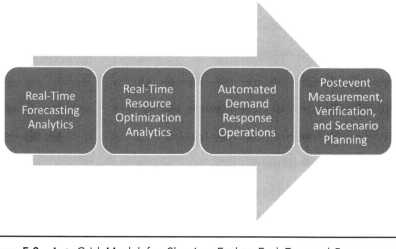

Real-Time Forecasting Analytics

Real-Time Resource Optimization Analytics

Automated Demand Response Operations

Postevent Measurement, Verification, and Scenario Planning

Figure 5.3 AutoGrid Model for Shaping End-to-End Demand-Response with Analytics. (Adapted from http://www.auto-grid.com/technology/our-first-application-droms, with permission.)

Very few vendors currently offer such a powerful approach to the use of analytics for C&I demand response, but the rapid application of analytics to solve utility problems combined with a cloud-based platform approach mark the beginning of a critical transition to end-to-end solutions. Entrepreneurial companies backed by advanced algorithmic models are setting this mark, but incumbents will certainly see the opportunity in moving their products beyond their traditional functional limits. Several cloud-based pilot programs are quickly leading to rollouts as utilities discover both the low-enablement and operating costs, as well as reliable load shed that is occurring with a unified and coherent approach to high-value C&I demand response.

5.2.4 Dynamic-Pricing Analytics

With the universal deployment of smart meters in the developed world a close reality, utilities, regulators, and customers are viewing dynamic pricing as a growing area of interest. Currently, most electricity customers pay a flat rate for every kilowatt-hour (kWh) they use, regardless of the time of day or the actual momentary cost to deliver that electricity. This disconnect in cost will only grow more difficult as renewable generation becomes a greater share of the energy mix, creating further volatility in the moment-to-moment costs of energy. The dynamic-pricing model changes the tariff system in such a way that a different price is charged at various times for the same amenity, but it's a price that reflects real costs. There are many challenges; utility rate-making is often a formal regulatory or government process for public utilities, and rate-making typically requires some attempt to set prices at just, reasonable, and nondiscriminatory levels.[10]

Dynamic-pricing advocates argue that customers will have more control over their energy costs. Detractors, on the other hand, worry that low-income customers—particularly the elderly and those with children and health problems, especially in areas with temperature extremes—will be unfairly impacted. It is questionable whether residential customers can hedge the risk of high prices like C&I customers can. To guarantee distributive justice, but also bring a tariff system that accurately reflects the imposed costs of providing electricity under different demand scenarios, customers will need to be able to change their behavior and shift load when prices are high. This implies that one of the keys to realizing dynamic pricing is the ability to offer products and services that guarantee instant demand-side flexibility and that reliably influence customer behaviors. This can only be accomplished adequately and at scale with data analytics that can help match supply and demand coupled with customer-side technologies that can automatically respond to price signals.

Influencing Behavior Requires a Relationship

In 2013, it was announced that, depending on the opponent and the game time, Toronto Maple Leafs fans are required to pay a higher ticket price for "better" games. The Maple Leafs' head office claims that it is simply market forces that are dictating prices, so it's entirely fair. What do fans think? Well, they don't seem to think it is a positive development. It makes them mad, actually. It doesn't seem to amount to more than a price-discrimination scheme, with a helping of inventory rationing. It seems simple at first: If fans want to see their team at its best, they should be willing to fork over more cash to do it. But, the system could backfire. As one commentator claims, "Reliance on strict demand-based pricing will tend to reduce the fan-team relationship to a series of cold economic exchanges."[11] Only time will tell the true impact of this decision.

Perhaps it seems intellectually dishonest to compare utility dynamic-pricing schemes with hockey ticket pricing. But, given the less-than-adoring

---

[10] J. P. Tomain and R. D. Cudahy (2004), *Energy Law in a Nutshell (Nutshell Series)*, West Group, Chapter 4, p. 392.

[11] Mike Lewis and Manish Tripathi (2013), "Why Sports Fans Hate Dynamic Pricing," Emory Sports Marketing Analytics (@sportsmktprof), *Emory University*. Retrieved from https://blogs.emory.edu/sportsmarketing/2013/08/21/why-sports-fans-hate-dynamic-pricing.

relationships that customers have with their electricity providers—and the fact that utilities must build trusting relationships with customers in order to implement demand-response and energy-efficiency programs—it is worth a moment of consideration. Yes, electricity isn't exactly an experience (but maybe it should be), and even if customers are used to paying a tariff for service, that is not the point. The point is that a dynamic-pricing paradigm shifts the focus from the receiving of an amenity to the fee itself.

Unhappy and ambivalent customers are a new issue for many electricity providers; back in the day, customers were just revenue generators. Now, the utility model is under threat of disintermediation and hollowing brought on by the availability of affordable microgeneration, community aggregation models, and competition for customer attention. By the time this book is published, fully expect to see the cable company moving into the position of energy provider. Thus, the real challenge is not how to algorithmically exact maximum revenue, but how to balance revenue generation with the all-important relationship with the customer. And ironically, it may be customer trust and engagement that will bring the acceptance of dynamic-pricing schemes in partnership with demand-response and efficiency applications, ultimately allowing utilities to implement effective pricing programs.

The Nuances of Pricing Schemes

It's not just utilities that are struggling to find ways to match revenues with delivering customer value. Consider this excerpt from a SAS Institute white paper on the topic of customer relationship management in the banking and financial management sector. Of particular interest for banks is understanding who their high-value customers are and what kind of banking products they should offer them. Traditional information, such as a credit score, is being analyzed with external data with great success:

> Many banks today are hoping to grow consumer and small business revenues . . . and also create "sticky" relationships that reduce attrition. . . . With high-performance analytics, the bank representative could assess the customer's current use of existing bank products and services along with associated profitability and combine that information with in-house propensity, credit scores and external data (such as outstanding loans and other financial relationships). . . . The overall value to the bank through the addition of high-performance analytics is that every customer interaction can be based on optimizing the price of new products for each customer in a way that increases retention, grows

revenue and improves the bank's profits while providing the optimal customer experience for each individual consumer or business client.[12]

Like the bank, the utility needs to maintain an increased understanding of customer behavior. Some analytic companies are beginning to make substantive incursions into this space to aid utilities in influencing consumer energy-use behaviors; although, because initial dynamic-pricing schemes are currently oriented toward C&I customers, residential solutions are lagging. Some of the analytical features that are proving fruitful in enabling new pricing programs include multidimensional analyses based on customer class, seasonality, and end-use characterizations, such as appliance type, integrated analyses to model the most effective ways to influence consumption choices, and elasticity modeling for price impact analysis. Visual analytics are an important part of creating a dynamic-pricing composite for the utility, and they provide visualization and simulation analytics of the impact of various pricing schemes on customers and their related load reduction trends.[13]

Effective analytical tools will help utilities gain the much-needed insight into which price schedules will drive the greatest influence on end-user consumption behaviors across appliance type, customer class, seasonality, and other factors. Simulation based on predictive and prescriptive models will help match energy-use behavior to various dynamic-pricing schemes and substantiate fairness for all customers by ensuring that they have the tools necessary to elastically respond to pricing schedules.

5.2.5 Revenue-Protection Analytics

Theft of electricity, euphemistically called "energy diversion," is a worldwide problem, resulting not only in major economic losses in terms of electric utility revenue, but in some developing countries, energy diversion creates a serious drain on already-strained infrastructure. Smart grid data analytics are playing a key role in identifying diversion scenarios that support revenue protection, including prosecution and collection. Energy diversion is accomplished with

---

[12] Ana Brown (February 16, 2012), "A Win-Win: Customer Relationship Dynamic Pricing," *The Knowledge Exchange,* SAS. Retrieved from http://www.sas.com/knowledge-exchange/business-analytics/innovation/a-win-win-customer-relationship-dynamic-pricing. Reprinted with permission.

[13] Space, Time, Insight: Dynamic Pricing Composite. Retrieved from http://www.spacetimeinsight.com/solutions/energy-and-utilities/smart-grid/dynamic-pricing-composite.php.

outright meter tampering, tapping into other premises, meter switching with low-consumption premises, or some form of meter bypass.

Methods for solving this costly problem are well suited to a wide variety of analytical applications. Accenture has developed a capability model for theft analytics based on grid infrastructure that correlates infrastructure maturity with analytic capabilities. The model begins with basic customer analytics and billing data and progresses toward smart grid feeder and transformer metrics. At the highest level of capability, the model emphasizes an aggregation of capabilities from the earlier phases of the model, emphasizing network visualizing and geospatial analytics.[14]

Like many comprehensive analytical approaches, the Accenture model demonstrates the cross-cutting nature of a fully realized approach to solving a specific utility problem, beginning with detecting anomalies in historic billing information and then correlating smart meter interval data, status and events, feeder analysis, and geospatial network visualization. Part of the value in constructing a capability continuum that is viewed through the lens of a powerful business need is the usefulness of such a model in developing a road map of solutions that can be aligned with other smart grid projects in the enterprise.

Specifically, the analytical techniques for utilities in detecting and closing energy-diversion incidents include analyzing data from customer information systems for similar classes and identifying usage anomalies and violations of predetermined thresholds and patterns based on a variety of characteristics and survey data. Smart meter data is especially useful because of the granularity of consumption data available to rapidly build a highly granular load profile that can be compared against other profiles and that accommodates seasonal or weather-related shifts. Third-party data can also be integrated into the data models, such as credit history, criminal history, and even social connections. Additionally, many of the techniques used to detect theft can be applied within the utility beyond diversion detection for distribution optimization, voltage and volt-ampere reactive (VAR) optimization, and fault location isolation and service restoration (FLISR) applications.

5.2.6 Breaking Down Functional Barriers

While we continue our discussion in further chapters with an overview of operational, customer, and cybersecurity analytics, it is likely clear that in

---

[14] Accenture (2011), "Achieving High Performance with Theft Analytics." Retrieved from http://www.accenture.com/SiteCollectionDocuments/PDF/Accenture-Achieving-High-Performance-with-Theft-Analytics.pdf.

the smart grid–enabled utility, data and functional silos will be naturally eliminated. As demonstrated, many of the enterprise applications that carry significant ROI depend on operational, business, and third-party data. The functional delineations that have served the utility so well as a largely project management–oriented operation are becoming more murky as the utility begins its transition toward brokering information through analytics for future business optimization.

Chapter Six

Operational Analytics

Apollo 10 firing room consultation among launch personnel. (*Source:* NASA[1])

6.1 Chapter Goal

The topic of analytics in the operational context is a broad and deep subject. This chapter sheds light on some of the most important driving issues

[1] Image retrieved from the public domain at http://grin.hq.nasa.gov/IMAGES/SMALL/GPN-2000-001849.jpg.

affecting how analytics are used in the control room and touches on some of the considerations for developing operational big-data analytics systems. These include the nature of control-room activities, the presentation of analytics for quick and effective decision-making, the automation of integrated distribution, the use of resiliency analytics, and the important role of standards.

6.2 Aligning the Forces for Improved Decision-Making

A concise definition of operational analytics will most surely lack precision. Defined in the negative, operational analytics is *not* about those precious aha moments or storytelling; it's about making better decisions in the moment. While there may be many beneficial side effects, such as customer satisfaction and optimization offered by a well-operating grid, improved decision-making is the primary reason for operational analytics. It is often assumed that only strategic analytics can have high economic impact on the business, but operational analytics run the gamut from low- to high-value decisions, with an aggregate of undeniably high-impact results.

As legendary business management consultant, educator, and teacher Peter Drucker said,

> Decisions are made at every level of the organization, beginning with individual professional contributors and frontline supervisors. [These] decisions are likely to have an impact throughout the company. Making good decisions is a crucial skill at every level.[2]

The relevant inference that we can draw from Drucker's observation is that every functional area in the organization requires adequate insight that can lead employees to make informed decisions. In an era of data-driven operations, this is how maximum productivity and profitability are achieved. An effective operational analytical model not only helps drive this required insight and understanding but also helps drive decision-making toward maximum high-value action. Operational functions especially require analytical capabilities that provide the tools necessary to make rapid decisions and use real-time data to solve critical problems in the moment.

Operational analytics are typically built on massive amounts of data with very low latency, and in many cases, they do not require human intervention. This is a function of automation often built and integrated on board grid devices. As

---

[2] Peter Drucker (2004), "What Makes an Effective Executive," *Harvard Business Review,* vol. 82, no. 6. Retrieved from http://hbswk.hbs.edu/archive/4208.html.

mentioned, since straight-through and onboard processing analytics are a highly specialized area, we will continue to focus on operational analytics that leverage the use of data mining, predictive analytics, optimization, and simulation.

6.3 The Opportunity for Insight

Providing insight into the grid from the substation to the customer meter is an excellent opportunity for utilities. The smart grid enables utilities to integrate instantaneous reporting to system operators from sensors installed at substations, on transformers, and from the sensing abilities within smart meters. Clearly, utilities now have instrumentation to help perform tasks on the grid that were simply not possible before, or, at the least, very difficult. By aggregating performance characteristics and information about load, operators are capable of understanding system load and utilization to ensure that assets remain within their operational parameters over time.

The management of intermittent renewable generation sources is another powerful use case that can provide a significant impact on reliability and coordination that goes far beyond existing supervisory control and data acquisition (SCADA) capabilities. This is a growing opportunity, as distributed generation technology brings energy storage, plug-in electric vehicles (PEVs), rooftop solar feed-in, and demand-response programs into the supply mix. Unfortunately, there is significant evidence that most utilities have not been able to analyze data beyond basic tasks of description, classification, and clustering to benefit from diagnosis and prediction. This is very likely due to a lack of experience in understanding the value of operational analytics, how to invest, how to establish return on investment (ROI), and how to work operational analytics into business strategy and planning.[3]

There is industrywide concern about the unknowns of operational analytics within the utility, but the technology required for these processes is actually quite well established and proven. Even so, utility organizations are well advised to start small and grow their efforts incrementally, investing not just in a big-bang technology but also in a managed rollout that prevents chaotic organizational shifts. Enterprise data also plays a role in operational analytics programs, especially for making predictions and extrapolating trends; in some cases, operational models rely directly on historical and business data, which are then executed against live data for real-time decision-making.

[3] BRIDGE Energy Group (September 17, 2013), "90% of Utilities Are Using Old Analytics Tools but Expecting New Results," *PRNewswire*. Retrieved from http://www.prnewswire.com/news-releases/90-of-utilities-are-using-old-analytics-tools-but-expecting-new-results-224062561.html.

6.3.1 Adaptive Models

In the operational domain, many of the most useful models employ adaptive methods that help improve the strength of predictive algorithms. With standard models, after deployment, they will continue to run until they are replaced with updated or refined models. Adaptive models continually self-tune based on the results they achieve. This means that the output itself is analyzed and based on a measure of success that's built into the model, and it adjusts to improve its results within the operational environment, allowing the system to adapt to emerging conditions.

Adaptive models are complex, and an explicit review of these modeling techniques is outside the scope of this book. However, it is helpful to understand that there are two types of adaptive models: those that take the content of the model into consideration and those that don't. Just because a model doesn't *understand* the content doesn't mean it can't be a useful and powerful tool. In fact, many early-warning systems comprise such adaptive models. If, in order to maintain ideal conditions, the model can identify when certain parameters fall out of bounds and can even correlate that data with other longitudinal types of data, the model can sound an alarm for action. What makes this approach adaptive is that it does not require a predefined workflow. Content-aware adaptive analytics (sometimes called semantic analytics) currently rely on the underlying data being tagged in some manner, but this will likely evolve as big-data processing technologies advance.

One of the aspects of adaptive analytics that makes it so well suited to the operational environment is that the data is described as entities—a location, a customer, or a service. Analytic models help improve confidence and have proved to be helpful for utility operators in developing confidence in their analytic models. For example, consider a predictive model that finds that a transformer that has been exposed to overstress events over a given period of time now has a 95 percent probability of failing within the next 30 days. However, if a certain detrimental event occurs (which can itself be predicted), there is a 50 percent greater chance that the transformer will fail in 15 days. When applied, the model uses refreshed data to adapt accordingly, providing much more powerful prescriptive insight.

6.4 Focus on Effectiveness

The utility operational context is really ideal for the application of analytics, and given a focus on effectiveness, it may be the best path forward to finding direct and tangible ROI from smart grid technology deployments. The authors at BeyeNETWORK wrote about operational analytics and noted that, despite the fact that not all operationally focused decisions have singularly high economic value, the sheer volume of these decisions can easily exceed in impact the value of one key strategic decision. In discussing how operational analytics can drive this value, the authors state, "Because operational decisions are repeated, they accumulate a large historical record of what works and what does not. Even when the historical data is missing, the repeatability of operational decisions lends itself to experimentation and testing to acquire data about what works and what does not."[4]

As described in Figure 6.1, operational decision-making is a constant cycle of analytical processing, risk and opportunity assessment for a particular set of actions or nonactions, decision-making and effect, assessment of the impact of the taken action, and subsequent adaptive tuning of the analytical model. However, the most powerful and effective models require historical data because their strength is dependent on working in the context of previous results.

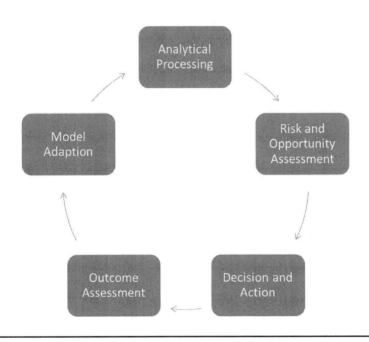

Figure 6.1   The Virtuous Cycle of Adaptive Analytics in an Operational Domain.

[4] BeyeNETWORK and Decision Management (2010), "Operational Analytics: Putting Analytics to Work in Operational Systems," report prepared for Oracle. Retrieved from http://www.oracle.com/us/products/applications/hyperion/operational-analytics-report-081829.pdf.

It is important to begin gathering entity-level data as it becomes available, and likely well before operational analytical programs have been well defined. One approach is to examine the low-value and low-complexity business decisions that can be automated, thereby freeing up the organization to focus on more-multifaceted and more-expensive decisions that require expert intervention. Begin by ensuring that this data is being collected and aggregated for use when the program begins to scale up. Properly designed, operational analytics translate into positive ROI directly by reducing the costs required to run the grid and indirectly by enabling better utilization of senior personnel skills and talents in an increasingly resource-constrained environment.

6.4.1 Visualizing the Grid

Real-time visualization of the grid is growing to be an important way to provide a coordinated response to leveraging powerful analytical models, which is discussed further in Chapter 12 but introduced here as a key to a fully realized operational analytics program. One excellent case study of the effectiveness of visualization for operational functions is the California independent system operator (CAISO, www.caiso.com) that manages the flow of 50,000 megawatts of electricity across the high-voltage transmission system in the state. These long-distance power lines make up approximately 80 percent of California's power grid in a state of nearly 39 million people with an economy comparable to Russia's.[5] The stakes are clearly very high, and reliable and safe electricity operations are demanded. Before grid modernization, CAISO was facing an operating environment where diagnostic support for the grid was not acceptable—or even available, in some cases. As part of smartening the grid in California, the ISO focused on providing situational intelligence to the dispatchers and operators in the control center through advanced visualization.

The program introduced the ability to gain real-time information of the grid through geospatial, visual feedback. Outfitting an entire 80-foot video wall, CAISO worked with analytics vendor Space-Time Insight to provide visual applications that correlate a wide variety of information, such as fire danger and crisis management data, the characteristics of various grid elements, and the weather impacts on distributed generation. Across the control room, varied actionable information is displayed and interacted with, including market, grid, and crisis intelligence; system planning information; and data required to

---

[5] *Bloomberg Businessweek* (2010), "California Retains Economy That Would Be 8th Largest." Retrieved from http://www.businessweek.com/ap/financialnews/D9JS1MLO0.htm.

successfully manage the intermittent renewables on the grid. Data tables to spot anomalies are very difficult for the operator to use without increasing fatigue; scanning visually across contours and colors provides feedback that a human being can quickly understand. CAISO reports that its ability to collaborate across functional areas has increased because the organization does not lose information that might be valuable to others, minimizing misunderstanding across multiple disciplines within the group and improving effectiveness.[6]

Unfortunately, a visual display does not necessarily mean that the value of an analytics application is improved. In fact, just the opposite can be true if the presentation of the data is junky and inappropriate. A well-designed presentation breaks down the barriers between how the operator thinks about solving its problems and accessing the information it needs to act. Intuitive systems can provide heightened situational awareness, increase the likelihood of an effective response, and enhance the prevention of accidents with catastrophic consequences. This is a very important issue, as described in an article about implementing data analytics situational-intelligence applications:

> Humans create mental models of the real world that help describe how things work; those models aid us in solving problems. Grid engineers use their own models, with system feedback, to make choices about the best possible course of action to keep the grid functioning. If the data presented does not align with that mental model, the engineer is left to continuously translate the information coming in, resulting in slower responses, fatigue, and a higher danger of making mistakes. A classic example of this phenomenon is the Three Mile Island accident, where the post-event inquiry concluded that the design of the control panel—specifically, a poorly designed and misunderstood light—was partly responsible for the disaster.[7]

Effective systems are not just analytically powerful and accurate. They also combine aspects of behavioral science and industrial engineering knowledge

6 Space-Time Insight (2011), "California ISO: Bringing State-of-the-Art to California's Grid." Retrieved from http://www.spacetimeinsight.com/pdf/Success_Story_Cal_ISO.pdf.

7 Carol L. Stimmel (2012), "Smart Grid Data Analytics in the Real World," *Smart Grid News*. Retrieved from http://www.smartgridnews.com/artman/publish/Delivery_Asset_Management/Smart-grid-data-analytics-in-the-real-world-4967.html#.Upp_8WST6xM. Reprinted with permission.

and experience for creating graphical display and interaction systems that are effective and usable by human beings working in high-stress environments. Data readiness is always a consideration in pushing fine-grained analytical models into the operational environment, but an ease of interpretation is what makes these models really useful.

6.5 Distributed Generation Operations: Managing the Mix-Up

Is the CAISO success with operational analytics enough to stave off anticipated grid problems caused by aggressive renewables mandates with intermittent renewables penetration approaching the 40 percent mark?[8] How about Germany, where the energy revolution (*Energiewende*) has brought a major shift toward green energy but also major problems related to integration issues, excess wind spillage, rising costs, and even a measured increase in year-over-year carbon dioxide emissions as coal plants are required to shore up generation in the cold months?[9] Certainly, wind and solar are becoming a significant part of the energy mix across the globe, and emissions standards are not slackening. But a lack of grid stability and flexibility to manage renewables is a grave concern; in fact, it is an impending crisis.

According to superintegrator IBM, there is a defined maturity model related to the operation and maintenance of renewables within the system that drives optimal functionality of these assets. This model is quite helpful in measuring the development of a renewables integration program against the business value that can be derived through optimal operations upon which the appropriate analytical capabilities rest. As it applies to analytics, within its end-to-end model, IBM includes monitoring, management, and optimization. Figure 6.2 describes just these elements.

Monitoring. Visibility is the first step to gaining control over what is often a very fragmented system of wind and solar generators. Analytical models contribute to dashboards, key performance indicators (KPIs) compliance, and real-

[8] Jesse Berst (2013), "WSJ Says What We're All Thinking: California Will Soon Have Grid Problems," *Smart Grid News*. Retrieved from http://www.smartgridnews.com/artman/publish/Business_Policy_Regulation/WSJ-says-what-we-re-all-thinking-CA-will-soon-have-grid-problems-5557.html#.UpqQtmST6xN.

[9] Spiegel Online International (2013), "High Costs and Errors of German Transition to Renewable Energy." Retrieved from http://www.spiegel.de/international/germany/high-costs-and-errors-of-german-transition-to-renewable-energy-a-920288.html.

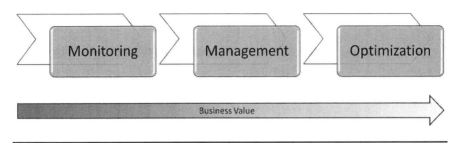

Figure 6.2 Business Model–Driven Approach to Optimizing Renewables Operations.

time monitoring capabilities. To even begin to achieve a useful monitoring state, utilities must also implement a complete data management solution, including acquisition, storage, processing, and presentation.

Management. For renewables operations and maintenance (O&M), management functions are largely dependent upon weather forecasting and high-quality prediction. Advanced analytics include models for numerical wind prediction to drive forecasted power outputs, enabling coordinated dispatch operations via linkages with conventional power sources, such as coal, gas, and storage. Integration issues for intermittent power sources are by far the most challenging obstacle for increasing the renewables mix; in fact, in some cases, they may actually result in system operators curtailing generated electricity that could otherwise serve the electrical grid. Additionally, analytics are also useful for predictive maintenance support to reduce overall downtime, report imminent loss of a system asset, and subsequently increase availability.

Optimization. As mentioned, generating capacity is not the issue with renewables; grid flexibility is. The greatest business value is achieved when analytical-driven tools serve to increase automation opportunities. Additionally, fully realized integration of intermittent energy brings opportunities for alternative business and economic models, such as transactive pricing and support for carbon and emissions trading. Fully optimized renewable-generating assets are critical to meeting social, political, and regulatory demands.[10]

Analytics are the keystone of the full industrialization of renewable energy worldwide. Predictive models are especially important to reducing the

[10] Rolf Gibbels and Matt Futch (2012), "Smarter Energy: Optimizing and Integrating Renewable Energy Resources," IBM: Thought Leadership White Paper. Retrieved from http://public.dhe.ibm.com/common/ssi/ecm/en/euw03067usen/EUW03067 USEN.PDF.

uncertainty caused by weather variations. At least in the short term, consumer demand and market conditions cannot flex to the granular fluctuations in weather that, through its impact on generation intermittency, can carry a gross impact on generating output. A viable and sustainable renewables program requires both accurate prediction models and automation to accommodate rapidly shifting power output from variable sources over time.

6.6 Grid Management

Analytics used for advanced, real-time distribution management are largely focused on optimization. These models are concerned with the functions within the grid's distribution network and may be used for performing state analysis; managing workforce; conducting fault location, isolation, and service restoration (FLISR); maintaining frequency and voltage levels; managing outages; and modeling and managing load. Other analytical models are quickly emerging out of operational necessity to help monitor and manage electric vehicles, distributed energy resources (DERs), and microgrids.

What are collectively called a distribution management system (DMS), various applications support control-room operators with monitoring and decision support for control of the electricity grid. In the United States, the real-time management found in today's DMSs evolved from outage management system (OMS) technologies, which comprehensively manage outage restoration across operations, crew management, and related customer support activities. In other parts of the world, pictures and papers were the heart of operations until SCADA systems allowed electronic control of operations. SCADA functionalities still play a role in the DMS topology, as do communication and remote-control capabilities.

Now, the smart grid brings a new level of complexity and capability to the DMS arena with distribution automation (DA) technologies such as reclosers, automated feeder switches, capacitor banks, and voltage regulators. As discussed, the relationship between DERs and DMSs is being defined even as these devices hit the network; new protection schemes and updated feeder configurations are required to prevent system disturbances, a meiotic term used to describe transformer explosions and blackouts.

Even the applications within the control room itself have tended toward point-to-point solutions that create both data and functional silos. In the era of the smart grid, we are going to see systems emerge that leverage a common network model and platform integration upon which traditional DMS—as well as OMS and SCADA—features are deployed. This unified approach brings flexibility, a simpler system, and the ability to quickly drive strategic initiatives into the action-driven operational zone. The benefits may be terrifically convincing, but making major changes to the DMS is what some in the industry call a "big bite." To imagine how difficult it is to upgrade command and control in the utility, consider how difficult it is to change the tires on your car—while it's speeding down the interstate. It's about the same thing.

It remains to be seen what predominant architectures will emerge to meet the grid operation requirements for integrated solutions that manage the entire smart grid life cycle, from DA to OMS to demand-response programs. Some vendors are calling for completely decentralized models that shift intelligence and control to the edges of the networks. That approach would necessitate predominantly onboard analytics of analytically driven control applications that would execute very close to the collection of sensor and other device data. Other architectures are much more of what could be described as hybrid. Specifically, this is a position of nonparticipation in the debate between distributed and centralized systems; instead, data is shared between the edge of the network and the central system. At this point, it is very unlikely that a utility would be comfortable adopting a totally distributed model; the evidenced conviction that these devices are truly capable of performing without human intervention is just not yet there.

In the cases of both the centralized and hybrid models, standards will be an important consideration for coherent information exchange. Working Group 14 of the Technical Committee 57 of the International Electrotechnical Commission (IEC TC 57 WG14) is developing IEC 61968, more accessibly known as the Common Information Model (CIM), which defines the interface between the major elements of the DMS. The standard specification can be found at the IEC smart grid standards website (http://www.iec.ch/smartgrid/standards). Given the sheer volume and variety of legacy and emerging types of data flowing off the smart grid, extending DMS capabilities to the utility without the benefit of communication standards is likely a foolish move. However, it remains to be seen if CIM is the answer to standardized data formats or if other definitions will provide better interoperability. For example, MultiSpeak, which is funded by the National Rural Electric Cooperative Association (NRECA), is already considered the de facto standard for interoperability. And more than 600 utilities in an estimated 15 countries are already using the format.[11] Also, importantly, the National Institute of Standards and Technology (NIST) has chosen MultiSpeak as a key standard for operations in the organization's conceptual model. There has also been some initiative to harmonize CIM and MultiSpeak and to enable interoperability between the two, a translation that allows end points that speak either CIM or MultiSpeak. Confusion abounds.

[11] MultiSpeak (2013). Retrieved from http://www.multispeak.org.

6.6.1 The Relationship Between Standards and Analytics

Analytic projects can be stressed and driven to failure by uncertainty. They can take too long to complete, and strategies can change. Standard communication models described by both CIM and MultiSpeak prevent redundancy in development efforts and facilitate the creation of unified, high-quality, reliable libraries that can become part of the analytics framework and, in platform service-based architectures, can be accessed through an application programming interface (API). Consider the example of an operational dashboard designed for two different consumers: a control-room operator and an executive overseeing operational outcomes. These two stakeholders require very different outcomes from their analytical software. The operator needs to be able to make decisions in the moment and to analyze a situation, she also needs the ability to rapidly drill down into root causes. The executive, on the other hand, is seeking to understand the story of what happened and why, and to analyze potential outcomes. Analytics help expose relationships between events, and even with the same data, different perspectives emerge. Standards at every level allow the utility to accelerate all efforts to modernize. And given the key role that analytics play in making sense of the state of the grid, standards are critical to the coordination of the innumerable disparate, point-to-point, legacy systems with digital sensors and other intelligent devices.

Utilities across the globe have invested heavily in advanced metering infrastructure (AMI) and smart grid technologies and continue to do so. However, these distribution devices are far from a full recognition of the technologies' value and potential benefits. At the same time, utilities are simultaneously in danger from underinvesting in analytics and overcommitting on smart grid performance. Distribution management, especially, does not lend itself to a project-based approach to analytics; though in the short term, progress can be made by building analytical models that rest on top of current data systems. Moving from the tactical phase of analytics for productivity toward more-strategic and more-predictive opportunities is when ROI will begin to reach its potential. Understanding the full capabilities of predictive and prescriptive analytics for the control room will bring the grid to express the vision of resiliency.

6.7 Resiliency Analytics

Making decisions to manage risk is the raison d'être for the utility control room. As DERs and unexpected phenomena such as significant weather events, both of which are impacting the grid at an increasing pace, are coupled with utilities' overall quest for efficiency and optimization, energy providers will need to take a proactive stance. Analytic tools that support both quantitative and qualitative techniques designed to manage the effect of reliability risks are emerging as one of the most important tools that the utility has at its disposal to address these unique challenges. Thus, in the domain of operations, analytics must incorporate well-developed and trustworthy risk models.

Developing effective risk analytics absolutely requires improving certain capabilities; these include the integration of models across many data sources, establishing linkages across functions, securely harvesting data in a manner that solves quality and integrity issues, acquiring skilled resources and expertise, and communicating insights effectively. Crisis management models, especially, raise the specter of resiliency within the grid and smooth restoration operations by providing damage estimation with improved situational awareness; they also manage work dispatch and predict time to restore. Strategic crisis models are certainly nothing new for the utility.

When Hurricane Sandy struck the northeastern United States on October 29, 2012, tens of millions of people were left without power, in some cases for weeks. Obviously, the grid cannot be completely hardened against flying debris and flooding, but resiliency measures are different than hardening in that they are designed to enable electric facilities to continue operating and to promote a rapid return to normal operations. An Institute of Electrical and Electronics Engineers (IEEE) *Spectrum* magazine assessment of the storm noted that smart grid design can be advantageous to resiliency, and included the following observation:

> When an electrical outage occurs, the smart grid's intelligent switches can detect a short circuit, block power flows to the affected area, communicate with other nearby switches, and then reroute power around the problem to keep as many customers energized as possible.[12]

Analytical modeling is the key to this resiliency. Because sensors and devices are able to report their status at microsecond intervals, it is possible to reconfigure the grid in subsecond ranges to restore power. However, any such reconfiguration must be able to support the load. One way to manage this is to tap DERs, such as battery storage and generators, as well as to kick off demand response, reduce voltage, and leverage available microgrid resources that allow certain parts of the grid to be isolated. To achieve this, the network must be modeled

---

[12] Nicolas C. Abi-Samra (2013), "One Year Later: Superstorm Sandy Underscores Need for a Resilient Grid," *IEEE Spectrum*. Retrieved from http://spectrum.ieee.org/energy/the-smarter-grid/one-year-later-superstorm-sandy-underscores-need-for-a-resilient-grid.

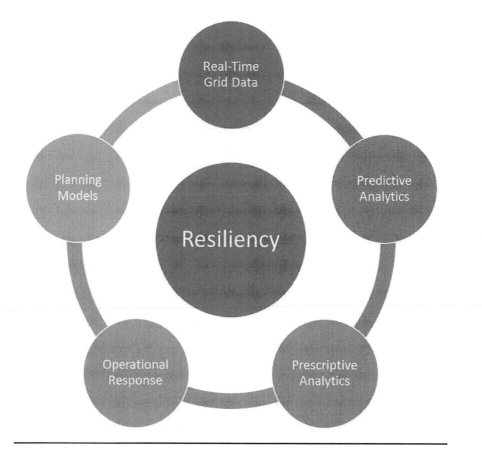

Figure 6.3   Creating a Resilient Grid Using Planning Models and Analytics.

and able to predict what is likely to happen next in the grid under a multitude of scenarios, and it must then be able to prescribe the best course of action.

Currently, most network models are used for planning purposes and not for operations. The Cooperative Research Network's (CRN's) description of the internal processes that lead to enhanced grid resiliency is interpreted in Figure 6.3. CRN project managers noted that improved accuracy and situational intelligence result from a better understanding of the network topology and lead to improved capabilities within operational functions.

6.8 Extracting Value from Operational Data Analytics

Once deployed, analytics can result in dramatic changes in how decisions are made as well as the rate at which they are executed. But getting there will take time and patience, since the investment return for analytics skitters around after the low-hanging fruit is picked. Automating reports and plugging in systems that drive productivity within operational functions will immediately provide an easily measured financial benefit.

The rest of the picture may not be so rosy, due to cultural and system barriers that make anything more than incremental improvement in using grid data difficult. For example, Oracle reported in 2013 that utilities are collecting a lot more data than they are using. This includes diagnostic flags, tamper events, voltage information, interval consumption data, outage information, and power-quality data. Nearly 40 percent of the data that 150 US and Canadian utilities are collecting on outages and 20 percent of the data they've collected on AMI is not being used in any manner.[13] This demonstrates an incredible missed opportunity.

It is easy to make the assumption that utilities are storing data for later; perhaps they're determined to put that data to good use as soon as they get their data management systems in place. Well, maybe. However, there is a concern that much of that data is actually being spilled on the floor and never even makes it to a data store. This is simple lack of preparedness. Utilities may have indeed created what Bob Meltcaf, co-inventor of Ethernet, has termed the "Enernet," but the capabilities and advantages are being wasted.[14] However, like the Internet before it, the advanced capabilities of the so-called Enernet will change many things about power delivery, including how energy is consumed. The operational domain, because of its characteristics of many low-value decisions with collective high impact, is the ideal first place to put analytics to work on the large amounts of very granular data that are now flowing from the grid.

Perhaps utilities have much to learn from the analytically based prowess exhibited by the telecommunications and financial industries. But there's a fundamental difference: In most cases, high-quality operations do not result in an increased sale of their product or a growing customer base. Instead, for the utility, analytical models must be focused on efficiency, cost reductions, and the careful management of challenges that can become very expensive and dangerous if they aren't contained. Operational analytics are required to meet the mandates for carbon management and greenhouse gas reduction, as well as the growing demand from customers to generate their own power, feed excess

---

[13] Jeff St. John (2013), "Smart Meters Must Better Integrate into Utility Operations," *Greentech Media.* Retrieved from http://www.greentechmedia.com/articles/read/Smart-Meters-Must-Better-Integrate-Into-Utility-Operations.

[14] Erik Palm (2009), "'Enernet'—A Smart-Grid Vision from a Net Tycoon," *Green Tech, CNET News.* Retrieved from http://news.cnet.com/8301-11128_3-10203683-54.html.

power in, and still rely on the macrogrid when they can't generate enough. Stakeholders who are designing analytical programs in utilities must realize that returns on big data investments are not easily comparable to other industries. The extraction of value from utility operations is much more about better decision-making in a rapidly changing and dynamic environment.

Chapter Seven

Customer Operations and Engagement Analytics

Apollo 11 astronauts swarmed by thousands in a Mexico City parade during the 1969 presidential goodwill tour. (*Source:* NASA[1])

7.1 Chapter Goal

The smart grid provides both direct and indirect advantages to customers, yet customer engagement may be the most important, albeit most difficult,

[1] Image retrieved from the public domain at http://grin.hq.nasa.gov/IMAGES/SMALL/GPN-2002-000016.jpg.

strategic imperative for the utility. In this chapter we explore some of the key drivers for integrated customer analytics that can serve both utility operations and customers within their homes. We focus on how to use analytics to increase residential customer lifetime value, improve satisfaction and trust, and incorporate the role of third parties in the customer relationship through the use of both well-known structured forms of data and data from emerging unstructured sources.

7.2 Increasing Customer Value

Most customer analytics systems are designed with the goal of predicting customer behaviors, largely as they relate to buying habits and lifestyle preferences. They are used predominantly in retail, finance, and customer relationship management (CRM) systems to calculate a dollar value for each household and to determine that household's value to a company, all with the objective of keeping valuable customers from switching to competitive products and services. Within the utility, until recently, gaining improved visibility into customers has not been a priority, simply because in many markets, customers are "captive" and cannot readily go buy electricity from a competitor. However, a perfect storm has formed in the electricity industry that requires the utility to better understand its customers: Rooftop solar has become more affordable, and the requirements for energy-efficiency and demand-response programs are growing. Even with the smart grid raising the specter for greater operational efficiencies, lowered costs of doing business, and improved billings and collections, the customer is becoming part of a participatory market and can no longer be ignored. Smart grids and smart meters are relied upon to provide the foundational data for developing powerful intelligence about customers that contributes to key utility functions and improves return on investment (ROI).

7.2.1 Customer Service

With the exception of competitive markets, the traditional role of the utility has been to deliver a single product to everyone; there has not been a high level of focus on customer service. In fact, the single most important concentration of the business has been meter-to-cash operations, which ensure that the utility is compensated for its service and that customers are billed a fair amount. In this situation, the utility seeks to drive down the costs of customer service while simultaneously maintaining reliable electricity delivery. However, with significant regulatory and political pressure to implement cleaner energy solutions and grid modernization activities, as well as the pressure of third-party competition, utilities must move in a much more service-oriented direction.

These services include managing distributed energy resources, supporting electric vehicles and home area networks, and finding a better way to engage the customer in energy-efficiency, demand-response, and conservation activities. Because of this, both regulated and deregulated utilities must now begin to understand their customers at a much more intimate level, and define and build authentic, trusted customer relationships. Traditional approaches simply do not meet these requirements.

In competitive markets especially, understanding how profitable a customer is to the business can be invaluable and is a key part of creating targeted service models. For example, a utility will want to retain a customer who has a low cost to serve. The best way to achieve this insight is to construct robust behavior models that help identify those customers whom the utility wants to retain for maximum profitability, and to concentrate strategic customer service programs that appeal directly to those consumers. Analytic models can be applied to raise awareness about what contributes to customer profitability, and those learnings can then be driven to frontline operations.

Overall, there are several key approaches that support enhanced utility customer service initiatives. They include the microsegmentation of customers, support for better targeting of and more-resonant marketing messages, customer sentiment analysis, and methodologies and approaches that can help create better customer engagement with utility programs.

7.2.2 Advanced Customer Segmentation

Because of the implementation of smart meters, utilities are moving toward using granular consumption data by combining it with other third-party information to develop enhanced customer segmentation models. For utilities that need to rapidly evolve with the changing landscape of energy delivery, the use of highly targeted segmentation models can help improve energy efficiency and peak-load-reduction outcomes. By better understanding residential and small commercial customer consumption behaviors, the utility is able to develop energy products and services that better target consumer needs, thereby increasing enrollment, ROI from grid modernization efforts, and customer satisfaction.

In the industry overall, the current utility understanding of customers is so poor that it is likely new business strategies based on a deeper understanding of the electricity consumer will quickly emerge. Unfortunately, the industry media has focused on helping utilities "survive" the smart grid evolution, and that dangerously distracts from identifying business models that will drive new products, services, and future prosperity. And in the future, leveraging the capabilities of the smart grid with the participation of the customer is key. Founder and chairman of the Smart Cities Council Jesse Berst underscores the peril in not understanding the customer: "[T]he smart grid's profound technology changes will be followed by profound business model changes. Unless a utility is working hard to profit and prosper from those changes, other companies will snatch them away."[2] As the technology of the grid changes, utilities must be prepared to serve as a bridge between regulatory and political mandates and customer needs. Either the utilities become optimizers or they will watch their business decline, and angry customers will move off-grid.

Predictive analytics are enabled by fusing utility data with a variety of third-party sources such as financial records, social media behavior, geographic information systems (GISs), and demographic data. Most utilities segment customers using an approach where historical data on transactions and other structured information on customer interactions are processed using rule-based systems. This approach is expensive and slow. Rules must be maintained and adapted and are usually quite simplistic. The influences that consumers are subjected to from news sources, events, entertainment, and especially social media are lost. In fact, they never even make it into the signal pool.

Alternatively, big data allows a more sophisticated approach by driving microsegmentation that creates a very precise view of a market. Instead of analysts creating stable rules, machine-learning techniques can be applied that make it possible to automatically generate multifactor rules with little human effort from the data itself. This allows the system to capture a large number of signals from both structured and unstructured data, and to adapt targeting approaches very quickly as consumer actions change. However, it is also important to remember that even the most sophisticated man–machine collaboration correlation requires human intuition to validate and apply, even when the rules are beyond human reasoning.

7.2.3 Sentiment Analysis

Sentiment analysis, also known as opinion mining, employs a combination of natural-language processing, text analysis, and computational linguistics to

[2] Jesse Berst (2012), "Why Utility CEOs Are Asking the Wrong Question (and What They Should Ask Instead)," *Smart Grid News*. Retrieved from http://www.smartgridnews.com/artman/publish/Business_Electronomics/Why-utility-CEOs-are-asking-the-wrong-question-and-what-they-should-ask-instead-4485.html?utm_medium=email&utm_source=Act-On+Software&utm_content=email&utm_campaign=Why utility CEOs a#.Up_fCWST6xM.

extract information from unstructured data. The goal of this form of analysis is to determine the attitude of the speaker or customer based on the context of the content, to understand online opinion, and to monitor reputation.

It has taken a few hard hits to drive utilities into considering the usefulness of maintaining currency and consideration for the importance of social media, blogs, and social networks, which are having a growing impact on smart grid implementations and programs. There have been three areas of major impact that have driven utilities to pay more attention to activities in the social sphere: health and safety concerns, privacy, and smart meter accuracy. The ability for formerly noninfluential people to gain credibility and efficacy for their opinions in social media has resulted in regulatory bodies demanding smart meter opt-out capabilities; it's also caused utilities to revise business models that once explicitly called for consumer benefits, motivated utility backpedaling on initiatives when technology solutions fell short, and prompted a growing overall social awareness of how electricity is delivered. Social media is also raising cultural-impact issues related to prosumer activity, affordability, and social responsibility.

These factors are ensuring that utilities begin incorporating consumer-related issues directly into business requirements and even technology designs. Consumer confidence indicators have long been relied upon by utilities as informers for marketing programs, but the relationship between these indicators and actual customer sentiment are not well understood. Consumer confidence surveys show that trust for utilities is dwindling, but these surveys do not tell us why and are rather coarse tools for watching trends and understanding the impact of utility initiatives. A 2013 Accenture survey reported that positive sentiment has been dropping: "Less than 25 percent of consumers trust their utilities. . . . Specifically, just 24 percent of consumers trust their utility to inform them of actions they can take to optimize energy consumption—dropping nine percent from 2012."[3] This is the lowest level of trust found since the survey was initiated in 2009.

This finding should be quite concerning for utilities that have a new level of dependence on customer satisfaction, and it underscores the need for utilities across the globe to get the basics of doing business with their customers right. That includes, more than ever, finding opportunities to maximize every energy consumer touchpoint, from social interactions to the bill. Part of this radical rethinking of how to create an "energy experience" for customers must include capturing subjective information, tracking trends, and employing this information for better marketing, detection of opportunities and threats, brand

[3] Barbara Vergetis Lundin (2013), "Consumer Trust in Utilities Lowest Since 2009," *Fierce Energy*. Retrieved from http://www.fierceenergy.com/story/consumer-trust-utilities-lowest-2009/2013-07-03#ixzz2ZEVWftnC.

protection, and ROI. There is a whole spectrum of tools, data sources, and analytical modalities that can be integrated to achieve these outcomes. The best early approach for utilities is an incremental one that contains up-front costs, keeping the adoption burden low for new sentiment analysis projects.

7.2.4 Revenue Collections

In almost every other industry, bad debt can be collected on by foreclosure or repossession. Utilities cannot do that; electricity that has been delivered and consumed is gone. Collection efforts can only happen in arrears. Predictive analytics help utilities see bad debt coming, allowing the utility to identify triggers and events as customers begin to show signs that they are going to have difficulty paying their electricity obligations. Once these triggers are identified, utilities can provide appropriate messaging to help avoid delinquency, using custom communications with strategies to help consumers conserve or use energy more efficiently, such as payment plans or low-income assistance plans. Reactive collections also benefit from predictive analytics by helping utilities optimize their collections strategies, reducing the costs of the collections process, and even ranking customers who are most likely to pay their debt.

Like every good analytics programs, breaking down both data and functional silos is paramount. A credits-and-collections suite of models can benefit from an extensive aggregation of data, including customer care data, consumption data, previous billing and payments data, grid data (such as outages and grid health that impact the customer), demographic and geographic data, satisfaction data, provisioning and repair data, marketing data, competitive data, adjacent market data, and a customer lifetime value assessment. A model might include customer behavior characteristics such as the number of automated teller machine (ATM) withdrawals in a month, bank balances, account arrears scores, and interest charges.

These models can be built by examining historical data and then attributing a score to each customer who establishes the probability of delinquency occurring. For collections analytics specifically, the best outcome is to develop a communications strategy that is automated and hooked into the models, so customers get the right message at the right time. The goal is to reduce the need for reactive collections that are increasingly ineffective with rising societal indebtedness, and to identify customers far earlier in the debt management life cycle for effective and appropriate customer engagement and intervention. Implementation of similar models in utilities has proved the application of predictive analytics for collections (or collection avoidance) to be quite efficient. In at least one case, Pitney Bowes reports that it found approximately 75 percent of the delinquents in the top 30 percent of the model population, a discovery that can speed up prediction cycles and response times for customer engagement.[4]

Revenue-collection analytics use a sophisticated approach that can help the utility drive customer loyalty and satisfaction, accelerate collections, and lower costs by supporting both short- and long-term strategies. A fully realized analytics-driven revenue-collection process can help continuously identify at-risk accounts and revenue bottlenecks, and intelligently prioritize efforts in a customer-specific, situation-appropriate manner.

7.2.5 Call Center Operations

With the emphasis on data flowing from the grid, the data generated by call center activities can be overlooked in the context of grid modernization. It is quite conceivable that, as the utility transgresses the smart meter into the home through in-home networking, automation, and interconnected nano-grids, the call center representative will be doing much more to support consumers, much in the same way that telecommunications companies adapted to support in-home networking for Internet services ("You sent me this modem, and I can't load a web page!"). The call center owns many important customer-related data streams, from accounting and claims adjudication to outage communications.

As utilities move from the simple delivery of a commodity product toward service orientation, they will need to implement more-sophisticated ways to measure the number of calls, their duration, average hold time, and resolution rates. Currently, the orientation of these measurements is agent- and efficiency-specific, but as more-technical queries and social media influence flow into the work stream, analytics are the key to providing real-time guidance that delivers greater quality of service to contain and drive out costs from the process.

Call center data analytics, like many powerful analytic models, bring together historical and real-time information to support decision analysis and improved product development. For example, analytics can help predict the root causes of customer dissatisfaction, understand the dynamics of expensive repeat callers, and distinguish revenue-related calls for special handling. Also, analytical models can be developed that can help improve agent responses with targeted training tools, even down to each agent. Some of these techniques relate to integrating sentiment analysis into contact center capabilities. For

---

[4] Pitney Bowes (2013), "Predictive Analytics + Customer Engagement = Bad-Debt Prevention," webcast presented by Energy Central. Retrieved from http://www.youtube .com/watch?v=wF3YlFc7GZk&feature=youtu.be.

example, identifying when problems or customer complaints are emerging gives the utility the opportunity to tackle them transparently and quickly.

Analytics are well poised to drive down operational costs even while providing richer, more-satisfying customer experiences. As mentioned, root-cause analysis is a key business motivator, and understanding customer pain points can even help teach an agent when deflection is the most appropriate response and when it will create a negative response. In fact, every customer action can be scored and quickly analyzed, improving the opportunity to efficiently turn around an at-risk interaction and improve overall customer satisfaction ratings. All these tools have one singular purpose that in many ways is similar to the goal of operational analytics—to drive the most appropriate, proactive, high-value action from the known information. The results of better decisions can be measured for their contribution to a positive ROI and position the utility to manage the inevitable rising complexity in the service-oriented, distributed organization.

7.2.6 Utility Communications

Due to the nature of the captive ratepayer found in many utility territories, customer churn is not a prevailing concern, with a low rate at approximately 9 percent in both emerging and mature markets.[5] However, declining revenue per meter is under downward pressure, and the changing utility business models and a new kind of churn are bringing the issues related to effective customer communication to the forefront. There are so many variations of energy regulation across the world that there are a number of specific reasons that any utility might want to improve utility communications. However, there are two driving issues that are key for the modernized utility: distributed generation and energy efficiency. These two issues require a level of service from the utility to manage its impacts, which include the physical and operational effects on the grid, declining energy consumption (and lower revenues) in developed nations, and the new technologies and diminishing costs of nano- and microgeneration that allow customers to "churn off the grid" altogether. In these instances, customers are accessing renewable sources of cheap energy, rendering the grid nothing more than a backup source of power.

Big data analytics are changing our ability to gain visibility into customer behaviors, but in many ways it has raised the difficulty level. For example, telecommunications used to seek to understand the duration of a call or to whom

[5] Astrid Bohe, Joon Seong Lee, Jim Perkins, and Jonathon Wright (2011), "Winning the Intensifying Battle for Customers," Accenture. Retrieved from http://www.accenture.com/SiteCollectionDocuments/PDF/Accenture-Communications-Next-generation-Customer-Analytics-Big-Data.pdf.

it was made; now it looks at which apps were used for texting, when Skype was used for a call, how Twitter and Facebook are incorporated into communications, and what story all of it tells about the life of a customer. With energy, it used to just be about keeping the lights on (KTLO); now the demands for reliability and quality are sky-high—every device needs charging, and the price is very high when commercial and industrial entities lose power.

Electricity is much less transparent than it has been in its most reliable years. We fight for outlets in airports, portable batteries are considered essential devices, and any outage is greeted with very little customer patience. Consider, for example, this tweet that @michaelsola made to his more than 1,500 Twitter followers about Baltimore Gas and Electric on June 2, 2013:

> Hey BG&E, who looses [sic] power at 630 on a Sunday morning in clear blue skies? don't make me fire up the generator. #fail #poweroutage

Since when is it cool to talk about backup power generators to over a thousand people? Maybe since February 2013, when a nationally televised power outage became the country's primary conversation. US-based Louisiana utility Entergy got the real-time drubbing of a lifetime after the Super Bowl outage that lasted for 30 minutes. Fast-thinking marketers jumped on the power outage, including such brands as Oreo, Tide, Audi, Volkswagen, even the Motel 6 chain, which chimed in with, "Next Year's Superbowl [sic] will be played at Motel 6. They'll leave the lights on."[6] All of sudden the social sphere lit up with messages about power surges, posts about relay-switch settings, jokes about carbon offsets, and speculation about the health of Entergy's grid assets.

The world of customer communication has changed for utilities. An online outage map and restoration estimations of 24 hours, followed by a very public argument about whose fault the outage was, are simply not going to assuage the digitally savvy customer anymore. Social media amusement was a huge embarrassment for New Orleans, which has been struggling since 2005's Hurricane Katrina, and changed the landscape of utility communications forever.

How Data Can Improve Communications

Currently, the utility relies on structured, transactional records such as customer interaction details that are usually of low volume and aged with low predictive value when they finally reach the modeling tools of the analysts and data

---

[6] Alex Kantrowitz (2013), "That Oreo Tweet Was Cool, but Is Real Time Marketing Worth the Hype?" *Forbes*. Retrieved from http://www.forbes.com/sites/alexkantrowitz/2013/02/06/that-oreo-tweet-was-cool-but-is-real-time-marketing-worth-the-hype.

scientists charged with improving customer communications. For real predictive value, analysts need access to unstructured forms of data, including social data, news and weather, in-home smart devices, and internal information such as outage and restoration activities delivered in real time.

Analytical tools—including web data extraction, text mining, and social media analytics that can detect sentiment and reputation—are frequently used in more-competitive industries such as retail and telecommunications. In the utility industry, these tools can be used to help identify key influencers that motivate the utility to craft timely, salient, and meaningful responses. It is emerging insights about customers and their communication needs that will drive the next generation of customer service.

As discussed previously, utilities are being compelled to move from project-based-operations companies that are focused on the delivery of the electricity commodity to service-oriented organizations that are uniquely defined by how their commodity is used by consumers, much like how we tailor our cell phone service to best fit our demand for mobile services. This will require improved agility in utilities' ability to respond to customer behaviors, actions, and market movements. Advanced customer microsegmentation also plays an important role in this agility by sending the right messages across geographies, genders, age groups, and other profiles. The most useful customer communication analytic tools will integrate measurement over results and the ability to help drive appropriate improvements.

Effective customer communication is fundamental to building trusted relationships with consumers. The examples demonstrate the rapidity of social communication in outage scenarios, establishing the absolute necessity for improved strategies for engaging with customers. Additionally, extreme weather events have shown the weaknesses in engagement strategies, and legislative and regulatory requirements are emerging that place specific requirements on utilities for communication during outage events. In fact, the New Jersey Board of Public Utilities is requiring the state's utilities to provide pre-event communications when possible, real-time outage maps, and estimated restoration times bound by strict guidelines.[7] Customer communication analytics are the key to developing and implementing integrated customer communications plans that are appropriate, timely, and effective at enhancing customer trust and satisfaction.

It is worth considering that customer communication, as a component of a customer analytics program overall, may be important to the industry in a

[7] iFactor Consulting (2013), "Three Important New Reasons Utilities Must Engage Customers," *Smart Grid News*. Retrieved from http://www.smartgridnews.com/artman/publish/Business_Consumer_Engagement/Three-important-new-reasons-utilities-MUST-engage-customers-6103.html#.Uq0XFmST6xM.

more pointed way as well. Shareholders show that a high value is placed on the management of information (Twitter and Facebook initial public offerings [IPOs] are demonstrable); thus, they are sure to expect the utility to learn how to monetize information assets for the benefit of customers. It will be demanded that utilities begin to think more strategically about improving customer value, with communications, energy products, and services as a hedge for future value.

> ### How to Start Building an Analytically Driven Customer Operations Strategy
>
> - Identify key customer service initiatives; understand who is doing the work, how work gets done, and the perceived benefits for customers
> - Design analytical models that help identify your customers at a microsegmentation level
> - Measure how your key initiatives are meeting the goals of your customers
> - Try to understand why your customers' desires (or their perceived desires) deviate from what you are providing
> - Strategize to tune your operations
> - Measure and adjust

7.3 What's in It for the Customer?

It has been quite clear since the inception of the smart grid undertaking that the key to success was for energy consumers to take a proactive role in their use of energy. Optimists will call this process of learning to engage the customer "evolutionary," but the more cynical among us will refer to the difficulties in winning utility customer trust as the industry's most difficult moment.

In the context of advanced metering, the notion of customer engagement seems to make the most sense, as these meters provide data that can be directly applied to nudging consumers toward conservation behaviors with better information and the opportunity for automation. However, customer engagement principles can also be applied more broadly with respect to dynamic-pricing programs, demand response, and even distribution automation as it applies to outage communication and restoration efforts. In any case, it is overwhelmingly clear that building customer acceptance and trust requires adaptation for the utility in a rapidly changing business environment. Clearly, this is not a one-size-fits-all effort.

The lack of successful customer engagement in the early phases of grid modernization was hopefully dismissed by utilities and regulators, but this failure to fully engage can no longer be ignored. Disenfranchisement among electricity customers continues to grow. With regard to the smart grid, their concerns are threefold: First, they are concerned about electromagnetic emissions that could have negative health effects; second, they have security and privacy concerns; and third, they don't understand the benefits of the smart grid and resent fees or charges that support its installation. This pushback can have far-reaching consequences and has caused regulators and utilities to take actions that are contrary to the goals of a unified smart grid, including adjusting technology approaches such as focusing on power-line carrier (PLC) instead of wireless, taking reactive defensive marketing approaches, and providing opt-outs from smart meter installations.

Neglecting to focus on the consumer directly in all aspects of strategic planning is risking future financial success; specifically, the utility that continues to target customers as captive ratepayers risks disintermediation. This will happen in much the same way that consumers stopped funding their savings accounts and began investing directly in the capital markets, cutting out the banking middleman. Without strong customer relationships, utilities too will be cut out, losing the opportunity to leverage their direct relationship with customers to protect both their core business and to sell enhanced services. Instead, companies like home security providers, cable companies, and telecommunications companies that can offer a compelling bundle of energy management services will prevail. These companies are already moving to fill the void for energy management services. As the incumbents, utilities have the advantage today, but they must move quickly to deepen their customer relationships to avoid squandering their natural leadership position.

7.3.1 Enhanced Billing and Customer-Facing Web Portals

The ability to influence future customer behavior is to effectively engage those customers. This is not an easy task for any organization and requires a variety of initiatives that map the promise of a brand to the customer journey. Enhanced paper bills have been the earliest and most successful way that utilities have found of capturing the attention of the consumer. These bills are personalized to customers and can directly help them take a proactive stance toward the issue of energy management; they can also provide learning opportunities and a rationale for taking efficiency and conservation action. Paper bills are also the primary channel to customers and can be used to integrate the utility into each customer's "energy journey," and as his or her level of engagement deepens, the utility can deliver new opportunities for learning, action, and adaptation.

The US-based Sacramento Municipal Utility District (SMUD) was the first utility out of the gate to conduct a fully controlled study of the impact of enhanced paper bills, called home energy reports (HERs). In a three-year pilot of the Opower HER program, which began in 2008, 35,000 residential customers received reports stating custom information, resulting in a remarkable estimated average savings yield of 2.9 percent per year, with an upward trend in savings.[8] The reports included three types of information: (1) how the customer's electricity consumption compared to his or her historical usage; (2) tips for reducing consumption based on custom information about the home; and (3) normative information that described how the customer's energy use compared to that of neighbors living in homes with similar characteristics.

The learnings from this study, especially the influence of normative behavioral messaging (peer comparisons), have led to new approaches. Some of these new tacks include customized web portals that leverage analytical models, such as dynamic tips based on results from questionnaires, populated on the fly, and combined with information from the customer's smart meter, weather data, physics-based building models, and other salient information such as the desire of the customer to maintain his or her comfort or to maximize cost savings. In a white paper on the topic, industry observer Bob Lockhart states:

> Effective customer engagement opens up new opportunities for utilities by creating a dialogue where before there had been none. Once the customer and utility have a way to talk, then there is a channel to introduce [new] initiatives. . . . As with any other form of engagement, the message on paper bills should be framed in terms that have meaning to the customer: why paperless billing is good for them (or the environment), how much they can save on their annual energy bills with dynamic rate plans, or how much easier life will be with the utility's web portal.[9]

HERs are now provided by several vendors globally to virtually millions of homes and have been touted as a revolutionary step forward for the utility that desires to create a new, more positive relationship with its customers. Using behavioral science and targeted messaging, vendors of HER products are reporting increased customer satisfaction, higher participation rates in other utility programs, growing acceptance of dynamic rates, and an increasing propensity to seek other ways to conserve energy, such as with in-home displays (IHDs). Further, this success is leading to new approaches to improving the

---

[8] Kevin Cooney (2011). "Evaluation Report: Opower SMUD Pilot Year 2," Navigant Consulting. Retrieved from http://opower.com/uploads/library/file/6/opower_smud_yr2_eval_report_-_final-1.pdf.

[9] B. Lockhart (2013). "Effective Customer Engagement: Utilities Must Speak Customers' Language," Opower. Retrieved from http://opower.com/uploads/library/file/24/Opower_WP_Effective_Customer_Engagement.pdf.pdf. Reprinted with permission.

effectiveness of other smart grid imperatives, such as demand response in near real time. These efforts are only possible as vendors are increasingly coming to market with advanced analytical platforms that are able to leverage a variety of big data sources such as weather stats, smart meter consumption data, real-time data from the home (for example, service-status and load-control data), and other disparate data, both structured and unstructured. However, the aggregate data is so voluminous that analytical models are absolutely required to not only carry out advanced-feedback and demand-response programs but also to design them to be engaging and effective.

7.3.2 Home Energy Management

The success of HERs has been a relief to utilities that rushed in early with new feedback technology projects that often—though not always—resulted in costly pilots and early missteps. As implemented, many home energy management system (HEMS) projects with IHDs were simply not effective ways of promoting response among customers, and it became an industry joke to measure the value of these devices with "mean time to kitchen drawer" (MTKD). MTKD rates the length of time it takes for the customer to stuff the display into the back of the junk drawer when it either runs out of batteries or becomes too ugly to maintain its pose on the kitchen counter. Yet, despite these early setbacks, HEMSs are beginning to catch on (though not the acronym) and they are now more captivatingly and collectively known as the "connected home."

The connected home is a powerful lever for the utility that must operate more efficiently, provide efficiency and conservation opportunities, and rely more on the management of end-use loads than ever before. Engaging devices such as Apple-like thermostats with powerful onboard adaptive analytics and within-reach price points are growing more attractive as these smart stuffs have begun to hit the "cheap suggestion list" in holiday gift guides.[10] It seems the utility sorely missed the boat with the IHD, but the hidden gem for utilities is in the growing consumer interest in the connected home. For example, instead of the utility having to provide and manage specially programmed hardware, such as a wireless thermostat capable of receiving setback commands from the utility, the new generation of connected devices allows the customer to choose from a wide range of options. The utility simply provides rebates or bill credits for every kilowatt-hour saved during peak times. Who or how these thermostats

---

[10] Katherine Tweed (2013), "7 Trends in Home Energy in 2013 and What They Mean for 2014," *Greentech Media*. Retrieved from http://www.greentechmedia.com/articles/read/7-trends-in-home-energy-in-2013-and-what-it-means-for-2014.

are managed to achieve their savings is of no interest to utilities because these devices simply signal the service provider that then manages the energy for the customer, while smart meters percolate along, providing accurate measurement and verification of participation.

Despite the advantages to the utility, this emerging market belies the notion that the utility will be in control of these services. In fact, the idea of routing information through the more ubiquitous broadband connection is a likely outcome as third-party service providers rush the market. For the well-positioned utility that has made progress in breaking down functional and data silos, this may not be a bad thing. These utilities have already been working to open their billing and other back-end systems to more easily integrate across the organization. The trajectory that the connected home has taken from its difficult start as a HEMS network is an early indicator that the energy services market is primed to break open and increasingly become attractive to new market entrants. In fact, utilities are beginning to understand that they can reach their goals with customers by focusing not on energy management but on comfort and convenience: better appliances, lighting controls, and automatic window shades that happen to drive efficiency goals. Utilities that can adapt to their role as either service provider or partner will benefit from the lower costs related to managing and installing hardware themselves and from a renewed focus on operational excellence for their customers.

7.3.3 Strategic Value

The implementation of customer-focused analytics can move the utility toward meeting strategic challenges. Unfortunately, it has been slow going when it comes to fulfilling the mission for customer engagement; often goals are lofty and vague at best. However, a truly comprehensive analytics platform allows the utility to understand how it's doing and to meet the dynamic forces of regulation, customer vicissitudes, and other shifts, while allowing tactical course correction against those factors. Analytics provide crucial insight into the success of engagement efforts by delivering:

- A pathway to increasing and sustained energy savings with information programs and the connected home
- Higher response rates to utility marketing campaigns, such as appliance rebates; appliance recycling; home audits; heating, ventilation, and air conditioning (HVAC); rebates; weatherization; and demand response
- Daily peak savings with greater efficiencies on the demand side of the meter
- Increased customer benevolence and satisfaction metrics

A fully integrated analytics platform incorporates both utility- and third-party-sourced data. This includes data that is native to utility operations, such as meter data values, program participation information, and weather and demographic data. Analytics not only enhance the effectiveness of customer-focused operations and engagement but also can provide the necessary measurement to dynamically adjust systems and services that evolve with customer engagement over time.

How to Start Engaging Customers with Analytics

- Identify and understand the value to your business if customers are engaged, invested, and motivated to have a relationship with your brand
- Align the vision for customer engagement, including how an authentic relationship becomes a corporate strategy
- Create a unified approach that transcends cultural divisions that can translate the utility's engagement goals into strategies, including marketing, communications, program management, and product management, as well as others that are empowered to make decisions and coordinate with one another
- Identify key performance indicators (KPIs)
- Work to align the information strategy for internal and external data management with the service strategy for engagement metrics
- Experiment with techniques and data sources to develop approaches incrementally
- Measure programs
- Adapt models and approaches

Chapter Eight

Analytics for Cybersecurity

Scientist analyzing the impact test results of a .22-caliber gun setup at NASA's Langley Research Center. (*Source:* NASA[1])

8.1 Chapter Goal

Cybersecurity is a major challenge for protecting the utility's critical infrastructure amid the growing population of critical cyber assets within the electric system. In this chapter, we explore the vulnerabilities, threats, and analytic approaches to responding to cyberwarfare against the utility, especially in the

[1] Image retrieved from the public domain at http://grin.hq.nasa.gov/IMAGES/SMALL/GPN-2000-001886.jpg.

context of the smart grid and the digital network that drives the modernized grid. Also discussed is the looming failure of traditional cybersecurity tactics in meeting the increased threat levels against the utility, as well as why a program of security analytics may be the best option for proactively and cost-effectively containing threats from the field, the enterprise, and even the physical plant.

8.2 Cybersecurity in the Utility Industry

There is very little clarity in the utility for understanding cybersecurity and cyberterrorism. Cultural and institutional fears range from the so-called digital Pearl Harbor to hacktivism, privacy violations, and Stuxnet-like sabotage. And anyone charged with critical infrastructure protection (CIP) who studies and seeks to prevent security breaches has the privilege of being a flat-out paranoiac (also known as someone who knows exactly what's going on). Yet, despite the popular media characterizations of cheese-puff-eating, Red Bull–drinking liberationists with bolt cutters, potential attackers may be nascent script kiddies, revenge seekers, organized criminals, or state-sponsored cyberwarriors.

Apart from these threat agents, it has been estimated that as much as 80 percent of information technology breaches are caused or assisted by people "inside" the enterprise. Either willingly or unwillingly (through social engineering) and with or without malice, security breaches are often enabled by a weakness created by people within the organization.[2] This doesn't mean that external cyber-attacks are overblown. In fact, the targeted advanced persistent threat (APT) is likely to be the most damaging to life and property through a grid security breach. It does mean, however, that assessing and managing risk with tools must include accounting for all attack vectors, including those from the field, the corporate network, and the physical plant.

8.2.1 The Threat Against Critical Infrastructure

In the past decade, the most secure systems in the world have been breached, including the National Aeronautics and Space Administration (NASA), the Space and Naval Warfare Systems Command (SPAWAR), the Federal Aviation Administration (FAA), the United States Air Force (USAF), and the White

---

[2] Major Barry R. Greene, CIO, G-6 Headquarters, New York Guard, as presented at the GovSec 2013 conference in Washington, DC, during the session "Critical Infrastructure Protection: The Enemy Within."

House. Clearly, utilities will never be immune to cyber, physical, or blended attacks. In August 2012, a power-line support tower was dismantled in the US state of Arkansas, and just weeks later, a substation was set on fire with a message carved into a control panel: "You should have expected U.S. [sic]"[3] And sometime around 2009, "Stuxnet," the digital worm designed to sabotage Iran's uranium enrichment program by damaging centrifuges, was unleashed, ultimately demonstrating the destructive power of the world's first widely known cyberweapon. In the United States in 2012, according to the US Department of Homeland Security Cyber Emergency Response Team (CERT), about *half of the reported critical infrastructure-related cyber incidents impacted the energy sector.* As the number of vulnerable points on the grid grows exponentially, this problem is only going to get worse.[4] Certainly, the notoriety and widespread damage that could occur from a successful attack on the grid are of interest to political hackers, lone wolves, and state-sponsored hacker gangs alike.

The nature of the intensifying problem is frighteningly straightforward: Because electric energy is generated and consumed almost instantaneously, system operators must continuously balance the generation and consumption of power. And the distribution of smart grid components requires a digital two-way communicating infrastructure to achieve this goal. Disruption of this infrastructure at a single point in the grid can have significant impact. As described in Figure 8.1, the smart grid adds layers of technology—such as data-transport and command-and-response applications—to the electricity infrastructure from network operations. With the increased use of smart devices enabled by computers, software, networks, and the enterprise, the risk of cyberthreats—both intentional and unintentional—has grown tremendously.

Not surprisingly, the United States Government Accountability Office (GAO) in congressional testimony in 2012 stated that malicious cyber activity targeting US computers and networks more than tripled between 2009 and 2011. The GAO states, "All of the onsite incident response engagements involved sophisticated threat actors who had successfully compromised and gained access to business networks."[5]

---

[3] Rod Kuckro (2013), "FBI Investigating Ark. Grid Attacks," *Utility Dive.* Retrieved from http://www.utilitydive.com/news/fbi-investigating-ark-grid-attacks/178875.

[4] Department of Homeland Security (June 2013), "Incident Response Activity on Internet-Facing Industrial Control Systems," *ICS-CERT Monitor.* Retrieved at http://ics-cert.us-cert.gov/sites/default/files/ICS-CERT_Monitor_April-June2013.pdf.

[5] Gregory Wilshusen and David Trimble (2012), "Challenges in Securing the Modernized Electricity Grid," *GAO Testimony Before the Subcommittee on Oversight and Investigations, Committee on Energy and Commerce, House of Representatives.* Retrieved from http://www.gao.gov/products/GAO-12-507T.

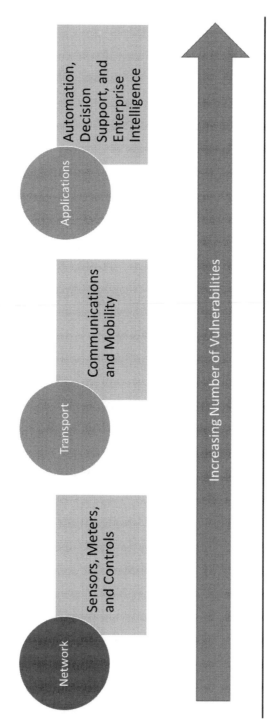

Figure 8.1 System Vulnerabilities Increase as the Layers of Complexity Expand.

8.2.2 How the Smart Grid Increases Risk

In the United States alone, it has been estimated that by the middle of the first decade of the 21st century, there will be more than 440 million vulnerable points on the grid.[6] This approximation includes smart meters, routers, smart building and home devices, and substation and distribution automation components. With the advent of the smart grid, the entire energy delivery process is digitized from generation to the point of consumption. Simply put, if any component on the grid can be communicated with, it can be exploited and controlled.

At its most rudimentary, the smart grid is a patchwork of supervisory control and data acquisition (SCADA) systems designed to operate in a distributed manner and bolted onto a digital control backbone that now is now managed through centralized operations. At its best, it is a well-coordinated system of networks, advanced devices, and industrial control systems. In either case, the grid has many digital touchpoints, and, consequently, weaknesses and vulnerabilities. What is more difficult to understand is the scope of the threat to these vulnerabilities. As Michael Assante, CEO of North American Electric Reliability Corporation (NERC), noted in a 2009 memo to industry stakeholders, "For cyber security, we must recognize the potential for simultaneous loss of assets and common modal failure in scale in identifying what needs to be protected. This is why protection planning requires additional, new thinking on top of sound operating and planning analysis."[7]

To help meet this anxiety, engineers, cybersecurity experts, federal security experts, and utility stakeholders perform simulations to grapple with the onslaught of computer viruses, exploding transformers and substations, and downed power lines. In late 2013, NERC ran a drill called GridEx II, which played out simulated loss of human life, denial-of-service attacks, and coordinated communication drills among power companies, local law enforcement, and cybersecurity control centers.[8] Called a necessary fire drill by its proponents, GridEx II has been labeled an exercise of academic interest by some industry watchers, suggesting that practices like this are doing very little to

---

[6] Darlene Storm (2010), "440 Million New Hackable Smart Grid Points," blog, *Computerworld*. Retrieved from http://blogs.computerworld.com/17120/400_million_new_hackable_smart_grid_points.

[7] Michael Assante (2009), "Critical Cyber Asset Identification," NERC. Retrieved from http://online.wsj.com/public/resources/documents/CIP-002-Identification-Letter-040609.pdf.

[8] Matthew L. Wald (2013), "Attack Ravages Power Grid. (Just a Test.)." *NYTimes.com*. Retrieved from http://www.nytimes.com/2013/11/15/us/coast-to-coast-simulating-onslaught-against-power-grid.html?_r=2&adxnnl=1&smid=tw-nytimes&adxnnlx=1384501226-AiLoSrpp3l0LspF4Ovg/hw&.

address the current threats. Perhaps the only clear point is that concrete steps and investment are required to begin addressing the most critical vulnerabilities in the system and face the reality of the threats that have been validated by a multitude of research studies, media exposés, and government assessments.

8.2.3 The Smart Grid as Opportunity for Dark Mischief

The consequences of a cyber-attack on the grid infrastructure include potentially massive and large-scale outages that could ravage the power grid. This is a reality that Joel Gordes, research director for the US Cyber Consequences Unit, describes as one that "[w]e are woefully unprepared for."[9] The following are the high-level scenarios for cyber-attack on the grid:

1. Reprogramming of critical electricity infrastructure components, resulting in major power delivery disruption
2. Theft of sensitive digital information used to mount later, more-coordinated attacks
3. Blended threats using a combination of hacking with a physical attack such as a fire or bombing

Due to the sheer number of points, their orientation in the outside plant (where they're hanging on people's homes, businesses, or nestled in their basements), and the subsequent ease of physical access, smart meters have drawn much attention for their ability to be hacked with just a modicum of skill. Some smart meters employ optical ports that allow utility technicians to diagnose problems in the field without disassembling the meter. These same ports have been used to reset the meters with an optical converter and downloadable software from the Internet to change consumption readings. This hack—or even the simple use of a magnet on the meter to disrupt recording during high-use times—can decrease a customer's bill by up to 75 percent. In one case, it was reported that upwards of 10 percent of the smart meters in one territory were tampered with and will continue to cost the utility over USD $400 million annually unless the situation is fully remedied. In this instance, the US Federal Bureau of Investigation (FBI) examined the fraud and concluded that former employees of the meter manufacturer and utility personnel were to blame,

[9] Patrick Kiger (2013), "'American Blackout': Four Major Real-Life Threats to the Electric Grid," blog, *National Geographic: The Great Energy Challenge.* Retrieved from http://energyblog.nationalgeographic.com/2013/10/25/american-blackout-four-major-real-life-threats-to-the-electric-grid.

charging a few hundred dollars to make the alterations for residential meters and several thousand for commercial devices.[10]

The safety of the data within the meters is also a concern. In the UK, this was one of the driving reasons that the Department of Energy & Climate Change delayed its 2013 rollout for more than a year. Part of the apprehension was that meter hacking can be used for more than just reducing consumption values; it can also be used for artificially inflating meter values or prices for personal revenge. More-catastrophic concerns related to hacking the entire nation's power grid via the smart meter network were also identified in the UK decision to delay its rollout. Acknowledging that the meter network is a strategic vulnerability makes it clear to governments and utilities that, despite the low-cost nature of digital meters, considerable investment is required to make both the devices and the systems more secure. This includes defending against hardware hacking and the mobile communication network through which the meter data is transmitted.[11]

Table 8.1 describes several of the common exploits that could be used within the context of the power grid. While a few approaches are enumerated, there are literally hundreds of these exploits and many frameworks that can be easily downloaded from the web to help even a novice attempt these attacks. It's impossible in a single book chapter to be exhaustive in a discussion as nuanced and complex as cybersecurity, but it's important to begin to appreciate the challenges in securing the grid and to provide a foundation for asking the right questions in developing initiatives and programs that are effective. Many of the exploits bear clever and arcane names, and their complete descriptions are out of scope. However, there are very comprehensive courses, books, experts, and other resources available to gain further knowledge to define and implement defensive approaches. The System Administration, Networking, and Security (SANS) Institute, a trusted cooperative research and education organization, is an excellent place to build foundational knowledge about information security principles.

8.3 The Role of Big Data Cybersecurity Analytics

Even where utilities are using big data analytics for various information and operational functions, the role of analytics for cybersecurity is not well realized.

---

[10] Brian Krebs (2012), "FBI: Smart Meter Hacks Likely to Spread," *Krebs on Security*. Retrieved from http://krebsonsecurity.com/2012/04/fbi-smart-meter-hacks-likely-to-spread.

[11] Zoe Kleinman (2013), "Smart Meters Need to Be Harder to Hack, Experts Say," *BBC News*. Retrieved from http://www.bbc.co.uk/news/technology-22608085.

Table 8.1 Description of Common Exploits on a Sample of Utility Systems

Utility System	Example Functions	Possible Exploits
Communications	Data transport, such as over broadband over power line (BPL), cellular, wireless, or satellite networks	Passive wiretapping Man-in-the-middle attacks Data modification Internet Protocol (IP) spoofing
Advanced components	Smart switches, storage devices, smart appliances, transformers	Routing attacks Denial-of-service attacks Node subversion Message corruption
Automated control systems	Monitoring and control systems such as voltage regulators and substation and distribution equipment	Botnets Zero-day exploits Modifications on controllers Spearfishing
Sensing and measurement	Smart meters and phasor measurement units (PMUs)	Wardriving Node capture Routing attacks Node subversion
Decision support	Operational applications to manage the electricity system	Structured Query Language (SQL) injection Buffer overflow Cross-site scripting Cross-site request forgery
Customer-facing systems	Web-based systems that provide account access to customers	SQL injection Cross-site scripting Denial-of-service attack Impersonation attacks

Yet, big data may be the lever that moves the industry from a reactive position on cybersecurity to one that allows for trusted prediction and a strategic posture.

As part of a holistic approach, cybersecurity analytics is just a piece of the puzzle for creating an in-depth defense of the grid. Imagine for a moment that your beloved grandmother lives alone in an apartment in the city, but she is very careful to lock her door for her safety and security, just as you ask her. However, the front doors of the apartment building are never lit or locked, the loading docks are unmonitored, and the windows over the fire escapes are unattended. Surely, this is not a comprehensive strategy for the personal welfare of your loved one—and we all know that hope is not a strategy. Building toward what cybersecurity experts call "defense in depth" is what we would expect: locking the building, hiring a doorman, installing security cameras on the loading docks, and developing policies that help the building residents work together for their mutual benefit. The same principles hold true for securing the grid:

authenticate, authorize, encrypt, detect policy violations, log events, and audit data. Cybersecurity analytics are a key part of building the required depth for maintaining a protected and resilient grid.

Critical systems are best secured by designing in security from the outset; that is by far the best approach to comprehensive risk management. However, where digital bolt-ons and legacy upgrades have occurred, it is crucial that the grid security architecture be augmented to provide the most secure operation. A program that constitutes a patchwork of reactive actions to address some discovered vulnerability is a program of denial for the role that utilities have in protecting civil society; it is also completely insufficient as cyberwarfare inevitably reaches the level of armed attack.

8.3.1 Predict and Protect

Incorporating cybersecurity analytics into the mix will begin to usher utilities out of a severe condition of vulnerability and address security requirements across the grid. There are several roles that analytical models can play that contribute to the overall protection and resiliency of the digital grid, including:

1. Gathering intelligence
2. Identifying industrial control system weaknesses and vulnerabilities
3. Quantifying identified threat levels and characteristics
4. Identifying real-time incidents
5. Predicting and preventing future incidents

Cybersecurity analytics has the potential to be a step-function improvement from traditional security models, which are largely passive defense systems and are fortresses only inasmuch as they resemble sandcastles on the beach. Current protections primarily focus on detection, but they ultimately fall to persistent attackers who avail themselves of an endless bevy of cheap exploits. In a quest for a more strategic and sustainable approach, big data intelligence strives to produce predictive results that give utility security analysts the ability to do more than just respond to attacks; they can actually stop them in their tracks. One approach to proactive cybersecurity is to become effective and efficient at recognizing attack patterns that represent threats. Big data analytics, by virtue of their ability to analyze massive volumes of data to drive actionable insights, are particularly well suited to detecting anomalous behavior on the grid.

A successful cybersecurity program in the utility will provide for full situational awareness across the grid and within the enterprise, deliver the ability to properly contextualize collected information, and enable the facility to quickly respond to and contain emerging threats. However, it must be said that one of the greatest impediments to fleshing out such a strategy is not political, organizational, or cultural; it is the lack of security features being built into smart grid devices. For example, there are smart meters available on the market that do not include such basic security features as event logging.[12] Without logging, it is nearly impossible to detect and analyze attacks, let alone prevent recurring threats. Not surprisingly, when the Industrial Control Systems Cyber Emergency Response Team (ICS-CERT) deployed several units in 2012—half of which were in the energy sector—to provide incident-response forensics, the team discovered that in many cases, conclusive analysis of the situation was impossible because of limited or nonexistent logging and the lack of other forensic data from the network.[13]

Big data analytics platforms combine security intelligence with powerful processing capabilities. The goal of such platforms is to provide repeatable pattern-detection algorithms with both structured and unstructured data sources, forensic capabilities, storage technologies, and enterprise integration functionality to identify both internal and external threats. And as with all advanced analytical solutions, these platforms will allow the utility to answer questions that have never been asked before. As Figure 8.2 describes, an integrated solution that provides closed-loop, continuous learning can furnish situational intelligence previously unavailable to security programs within the utility, both for information and operational concerns.

There are a variety of capabilities that are offered by a big data approach to cybersecurity, including:

1. The ability to detect anomalies by identifying correlations across disparate data
2. Real-time query capabilities
3. Visualization and exploratory tools
4. Postincident forensics that help improve detection algorithms

To successfully realize these capabilities in practice, a particular technical challenge needs to be addressed: the lack of a known baseline from which risky situations can be derived. To overcome this, the utility must be capable

Figure 8.2   Situational Intelligence from Collection to Response.

---

[12] Wilshusen [5].
[13] Department of Homeland Security [4].

of analyzing many months of network traffic, device information and parameters, communication characteristics, and user behavior to understand both the device and human linkages that make up the system. This is especially crucial for identifying more-random or more-infrequent forms of activity on the network that may exist in high-volume, high-velocity data traffic as found in many of the grid's command-and-control systems.

8.3.2 Cybersecurity Applications

Lessons from other industries, especially large-scale financial systems, are instructive and low risk in terms of cost to initial deployment. Continuous monitoring is a process that is especially efficient and effective at addressing operational systems. Born from traditional auditing processes, continuous-monitoring systems can be a key piece of the puzzle in identifying problems or weaknesses. Mark Nigrini, in his book *Forensic Analytics: Methods and Techniques for Forensic Accounting Investigations*, focuses on the discovery of anomalies in transactional systems, but two of his approaches are informative for the utility system as well.

First, he describes the approach of parallel scanning, which uses descriptive analytics for one period of time and compares the information to data from a prior period. Large differences indicate a signal for an anomaly. This can be done in a matter of minutes and can suggest that a system is experiencing an out-of-bounds condition. Secondly, risk scoring can be used as a method of prediction that, based on predetermined characteristics, assigns a risk value. A high score can help the utility orient and prioritize the devices or areas of operation that require high-priority attention.[14]

Big data cybersecurity applications must be able to manage and process millions of events per second with microsecond latency from both traditional and nontraditional sources of data. They must also be levers across multiple outputs, including:

1. Reporting
2. Visualization and exploration
3. Predictive analytics
4. Content analytics
5. Energy industry–specific applications

These platforms support the continuous ingestion and analysis of data, with as little impact as possible on the underlying infrastructure by using scheduled

---

[14] Mark Nigrini (June 2011), *Forensic Analytics: Methods and Techniques for Forensic Accounting Investigations*, John Wiley & Sons Inc., Hoboken, New Jersey.

polling and streaming of data sources. Scale is a challenge, and some solutions will fuse data elements to produce efficiencies and decrease latency in communication networks.

8.3.3 Proactive Approaches

The scope of the threat is beginning to drive innovation that works to actively and more directly deter the movements of attackers. CrowdStrike's co-founder and chief technology officer (CTO) Dmitri Alperovitch has coined this approach known variously as "active defense" or "proactive response." Alperovitch notes that current passive security models will continually drive up costs without a coincident level of effectiveness. Turning the tables on the adversary, active defense attempts to drive up the costs and risks associated with their hackers' activities. Instead of focusing on the often mutable characteristics of each discrete attack, the active defender focuses on identifying the mission of the attack and the tradecraft employed by the intruder.

Once the mission of an attack is understood, passive defense strategies are augmented by deceiving, containing, tying up resources, and creating confusion that increases the costs to the attacker and allows defenders to both isolate the attack and continue to collect additional intelligence.[15] Information collected about unique attackers can serve much like a fingerprint, allowing joint action with other utilities and government agencies to prosecute threat actors. This approach amplifies the efforts of cybersecurity to exclusively identify and predict patterns based on attack vectors, and it exploits characteristics to greatly improve the forward-looking stance of the utility defense model.

8.3.4 Global Action for Coordinated Cybersecurity

Expanding awareness of cybersecurity threats has brought attention from regulators and governments striving to produce suitable laws and standards within the utility. Industry standards for cybersecurity have been most prolific in North America, particularly the United States, Canada, and a portion of Mexico; though it is a global concern, and the lack of resolution has slowed further smart grid deployment in some regions. Several initiatives are serving to progress cybersecurity, but the most important first step in developing a comprehensive program is to not only understand but to also engage and comply with the NERC CIP standards.

---

[15] Dmitri Alperovitch (n.d.), "What Is Active Defense?" Retrieved from http://www.crowdstrike.com/active-defense/index.html.

NERC is the electricity sector's coordinator for CIP, and the firm has invested heavily in standards development, compliance enforcement, and the provision of extensive technical material and subject-matter expertise. The NERC CIP standards are the only mandatory cybersecurity standards in place to address the security and reliability of the electricity grid. The nine standards include mandates for incident reporting, authorization protocols, minimum security management controls, and disaster recovery. It is indisputable that the efforts of NERC CIP have reduced risk and improved the security posture of North American bulk electricity systems. However, it is equally indisputable that it is impossible to address every security risk, and this is precisely why the opportunity for predictive cybersecurity analytics may hold such profound value for more-advanced security controls.

Collaborative approaches to cybersecurity are taking hold and are an acknowledgment of the hugely complex problem that cybersecurity raises for CIP. The National Institute of Standards (NIST) and the Edison Electric Institute (EEI) are both working to improve responses to grid threats and vulnerabilities. To that end, NIST has founded the National Cybersecurity Center of Excellence (NCCoE), which brings together researchers, users, and vendors to perform targeted testing to improve cybersecurity outcomes. Focused solely on the mandate to deliver reliable power, EEI develops principles and provides clarity in the field of CIP. Among developing legislation across the globe, there are also several other public and private partnerships designed to strengthen the cybersecurity posture of the electricity sector. Information sharing will only improve the ability to develop more-effective practices and approaches to protect grid assets from all levels of attack.

Collaboration is particularly important among various stakeholders, because one way to dramatically improve the effectiveness of cybersecurity analytics is for utilities to share information. Despite a few steps forward in this direction, there is currently a lack of effective mechanisms to disclose vulnerabilities, threats, best practices, and actual incidents. This is likely due to the natural avoidance of publicizing attacks against the utility, but it carries the negative impacts of stifling informed corrective action, future defense, and maximized cybersecurity investments. Progress is being made, however. In 2013, the US House of Representatives passed the National Cybersecurity and Critical Infrastructure Protection Act (NCCIP Act), which is designed to help facilitate the real-time sharing of threat information across critical infrastructure sectors.[16]

[16] Jones, S. (2014), "National Cyber Security and Critical Infrastructure Protection Act Passed," *Incident Communications Solutions*. Retrieved March 13, 2014, from http://incidentcommunications.com/blog/national-cyber-security-and-critical-infrastructure-protection-act-passed?utm_source=Plazabridge&utm_medium=email&utm_campaign=ICSMarchInsights%3ANewsletter.

On a global level, this much-needed collaboration is also beginning to solidify. In 2013, at the Group of Eight summit, a finalized agreement was announced between US President Barack Obama and Russian President Vladimir Putin that introduces measures in the cyberdomain, including information exchange and crisis communication. China and the US also made progress through the creation of a working group on cybersecurity issues—a major step forward, given the mutual accusations of cyberwarfare. Both of these agreements were the reinforcements for the generation of a groundbreaking UN report that proposes international cooperation measures, confidence-building, and improvements for protecting critical ICT infrastructures.[17]

8.3.5 The Changing Landscape of Risk

It's a truism that absolute cybersecurity is not an attainable goal, and the grid will never be gold-plated either. It's also a fact that, despite the critical nature of the grid, cybersecurity must always be approached pragmatically, through a combination of assessment of perceived risk and the costs of security. Utilities must fully account for the risk of losing various grid functions; the impact of that loss; and how they can protect, detect, and respond to various cyberattacks. However, utility stakeholders must also understand that even the hint of a cyberthreat in a cross-cutting network means that multiple assets can be remotely attacked at once. Unlike reliability risks prior to the smart grid that could be accounted for in operating assumptions and planning exercises, as they are largely probabilistic failures, the digital-communicating nature of the smart grid requires a broad perspective and a shift in risk analysis.

The traditional perspective of risk is a simple algebraic equation that multiplies the probability of an occurrence with a measure of impact. To make more-informed decisions related to cybersecurity-related risks, the FBI recommends an expansion on this equation. Specifically:

$$risk = threat \times vulnerability \times consequence^{18}$$

According to the FBI, each factor is crucial because it moves the organization beyond a rigid focus on threat vectors and actors. As FBI experts point out,

[17] Detlev Wolter (2013), "The UN Takes a Big Step Forward on Cybersecurity," *Arms Control Association*. Retrieved from http://www.armscontrol.org/act/2013_09/The-UN-Takes-a-Big-Step-Forward-on-Cybersecurity.

[18] Ben Bain (2010), "FBI Outlines Three Components of Cyber-Risk." *FCW*. Retrieved from http://fcw.com/articles/2010/02/24/web-afcea-cyber-panel.aspx.

the risk model is quite useful when a strategic viewpoint is needed, and it helps define goals by pushing any of the variables in the equation toward zero, which would close the risk.

Many have noted that a literal interpretation of this equation is complete nonsense; despite its relevance to understanding the probability of an incident occurring, it cannot signify absolute risk because the variables do not carry units of measurement. Do not for a moment be tempted to populate a massive spreadsheet cluttered with assets to determine threat-ranking outputs that help you design a cybersecurity program. This would be just as useful as multiplying "purple" × "meat thermometer" × "lamp" (to severely overstate the case) to calculate risk. Risk-profiling of this nature may be a useful management tool to aid decision-making, but it quickly becomes absurd for use in developing a defensive strategy.

Cybersecurity analytic approaches that provide real predictive value to the utility more closely resemble models of complex systems. By viewing the network as a system of relationships, we can understand that the couplings within the smart grid are more similar to the human brain and are thus not well suited to linear analysis. Complex systems theory shows how these relationships within a system give rise to a form of collective behavior that in many ways is defined by its relationship with its environment. Making predictions about the behavior of the grid under attack conditions is what we need to understand in order to move from a reactive to a proactive posture. As the Nobel Prize–winning economist and philosopher Friedrich Hayek observed, complex systems' behavior is best predicted through modeling and an understanding of its patterns rather than precise predictions. Big data cybersecurity analytics can provide just that.

Key Considerations for Establishing Big Data Cybersecurity Analytics in the Utility

- Identify information security issues and evaluate the role of big data analytics
- Seek to resolve deficiencies in cyber-readiness, including in professional staff, governance, and information technologies
- Work to move from a defensive, reactive posture to a proactive system that accounts for the nonlinear characteristics of the smart grid
- Consider the roles of collection, storage, and processing apart from desired analytics and workflows
- Enable data and information sharing with other entities, including other utilities and cybersecurity entities
- Create a small pilot opportunity to prove the value of big data analytics to the role of cybersecurity
- Develop use cases that support business and operational vulnerability and threat detection

Section Three

Implementing Data Analytics Programs for Sustained Change

Chapter Nine

Sourcing Data

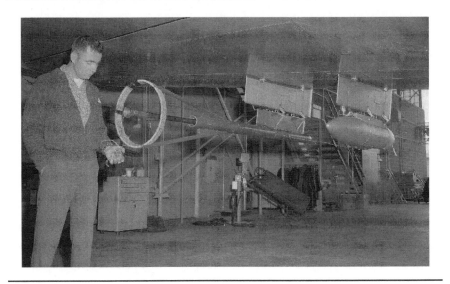

This B-5713 airplane is designed to collect radiation from an experiment mounted on its wings. (*Source:* NASA[1])

9.1 Chapter Goal

Preparing to successfully introduce a big data analytics program into the utility is predicated on a deep understanding of available and desired data sources as well as the business value of that data. A variety of data sources are discussed in

[1] Image retrieved from the public domain at http://d3.static.dvidshub.net/media/thumbs/photos/1302/861971/450x360_q75.jpg.

terms of how their characteristics bring value to the optimization of the utility, from both an operational and a business perspective. Data fusion, the implications on privacy, and the value of collaboration among utilities is also evaluated. We'll cover the devices on the grid that provide situational intelligence, how aggregated data can drive new insights, and the complexities of data-fusion models used to create those aggregations.

9.2 Sourcing the Data

Mark Twain is thought to have said, "The secret of getting ahead is getting started." So, where do we begin when it comes to approaching big data projects for analytics in the utility? Often, when faced with the onslaught of information flowing off the grid, all utility stakeholders will ask, "Where are we going to put all this data?" In fact, most consultants working with utilities on early projects will begin with the repository—specifically, capturing and organizing data. It's reasonable; after all, we don't always know the value of the various forms of data even once they're in the system, and it is certainly fair that we don't know at all the scope of the questions and answers that will be enabled by the data (or that may materialize later). In fact, unlike typical data warehouse projects that anticipate how the data will be analyzed and that categorize the information at the point of entry in preparation for specific analyses, big data projects are best served by massive data stores where the information can be easily retrieved in myriad ways by many analytical applications.

Even with the need for extensive infrastructure and tools, big data analytics are at the very core a business challenge, not a technology problem. Focusing on data management issues right up front is premature and can cause expensive missteps. Focusing primarily on the technology problems to "get data" instead of solving issues and finding new opportunities by using analytics is putting things out of order. It doesn't make sense to create a haystack and then go looking for a needle; meaning, it is not always necessary to collect lots and lots of data just because it might be useful later. That's called hoarding, and it's a fear-based response to not understanding the problem domain. Even with the rapidly diminishing costs of hardware and the low-cost scalability of new big data systems (thanks to open-source entrepreneurs), the expenses of operation, application development, and skilled management are not scaling nearly so efficiently. The inconvenient truth of big data analytics is that the costs of processing, querying, managing, and trying to extract value from stale data may be slowing things down.

The challenge is not to start hiding data in the data closet, but to understand and distill what is useful within the flowing river of data, and to minimize the rest. Thus, every enterprise must determine what can be extracted from the raw data, and companies should try to build an understanding of future value that leads to a rational architecture. Tools and techniques are evolving quickly and largely meeting the needs of big data analytics from a technical perspective, but for the goals of the utility, we must closely examine how to best exploit the big data opportunity. A good starting point is to bring the organization together to ask, "What kind of problems do we have in the company that we think data will help us solve?" quickly followed by "Do we have the data that will help solve those problems?" "How do we get the data we need?" and "How fast do we need to get there?"

To begin answering those questions, the utility must take stock of its assets to understand what it has, what the possibilities are, and what additional data is necessary to generate the necessary benefit and establish rapid and acceptable return on investment (ROI) from what will ultimately comprise a cross-cutting effort. Determining what data will feed the requirements is a more difficult thought exercise, but it's impossible until the existing business value of the grid and enterprise data is understood.

In Chapter 3, we examined the functional characteristics of grid data classes that are in use at the utility (Table 3.1). To recap, these classes are telemetry, oscillographic, consumption data, asynchronous event messages, and metadata. Additionally, customer, enterprise, historical, and third-party data must be accounted for. The business value of each of these data classes, however, is variable depending on how it is used by the utility. With data analytics, a single data class may have value, but when it's combined and analyzed with other classes, it can support many other surprising business needs. Understanding the underlying data is the key to aligning the subsequent architectural and technology decisions that must be made with solving high-value present and future business needs.

9.2.1 Smart Meters

Smart meters are often believed to be primarily consumption devices, mostly because the smart device has replaced scalar meters for meter-to-cash operations. However, despite the lack of global specifications about the capabilities that a smart meter must meet in order to be considered a smart meter, most of these devices will provide power-quality measurements, such as line voltage, current, and frequency, above and beyond the clocking of interval data. With these enhanced capabilities, smart meters can play an unexpected role in grid troubleshooting, maintenance, load planning, and—in the case of smart meters that are designed to carry signals to in-home devices—demand response.

Meter data collection has largely been the purview of meter data management system (MDMS) vendors, and many vendors are pushing the traditional confines of MDMS to provide analytics—most often outage-notification and revenue-protection analytics. MDMS has been a natural starting point for smart meter data analytics because it is already a working repository for consumption data and is often designed to interface directly with billing, maintenance, forecasting, and customer service systems.

Meter-to-cash operations will always be a utility business function and certainly one of the most valued and protected functions in the enterprise. Some industry leaders have demonstrated trepidation and concern over anything that could suggest tampering with this core function. Yet, there are several business problems that smart meter data, through the use of analytics, can help address powerfully and effectively:

- Improve the uptake of demand-side management (DSM) programs
- Boost customer satisfaction ratings with better outage responsiveness and communication
- Reduce revenue loss through better identification of theft
- Improve load forecasting
- Enable the provision of new energy services
- Develop new rate plans and services

Smart meter data analytics are well poised to lead the way to improved relationships with energy customers, notably by solving relevant problems for consumers. Analytics can also drive the profitability of the utility itself by helping to identify failing transformers and improving demand forecasts, revenue protection, and overall operating efficiencies. Leveraging just the data stored in the MDMS database, utilities can make significant gains by analyzing the meter events and readings. Truly, this is an excellent starting point for bringing analytics to the utility enterprise, particularly given the advantage of immediate ROI achieved by augmenting a business function that enduring.

Not surprisingly, the need for meter data does not stop at the MDMS boundary, despite the advantages of isolating the meter-to-cash operation. Depending on the overall architecture of the system, while some analytic processes may reside directly in the MDMS database, the data can be shared into a greater analytics platform. This can be accomplished using either a scheduled extract, transform, load (ETL) process (though this solution will strain under billions of transactions) or a real-time message bus that carries the data to the analytics platform, which may reside in the enterprise or in the cloud. To support functions such as theft detection, outage restoration, mobile workforce management, voltage/volt-ampere reactive (VAR) management, and predictive load modeling, it is crucial that the data from the meter get into the system expeditiously. This new phase of innovation will occur as utilities become more comfortable with the stability of their meter-to-cash functionality. It will also materialize as utilities begin to merge their meter data into a gridwide platform that feeds various applications, including visualization and geographic information systems (GISs), outage management systems (OMSs), distribution management systems (DMSs), and demand-response management systems (DRMSs), but especially for the broader analytics effort, where a deeper understanding of grid and customer behaviors will be realized.

9.2.2 Sensors

While smart meters can and do serve as sensors, other network data is collected from sensors along the transformers, power lines, voltage detection devices, and DSM equipment on the load side of the meter. All this data is key in addressing business and operational issues. In addition to sensors, other monitoring equipment provides a complete view of the state of the grid with information about overall operating parameters. These sensors may be state-of-the-art digital nodes or retrofits on legacy equipment, including clamp-on devices, and they maintain wireless communication. Many smart grid sensors are composed of a transducer that converts physical forms of information to an electronic signature, a central processing unit (CPU) for onboard processing, and a communications module that transmits the information over a high-speed network or wirelessly through a transceiver. Of course, in a distributed environment, not all sensors are designed to return data to the utility, as they employ circular or first-in, first-out (FIFO) buffers and are built to automatically respond to certain inputs. Sensor data that does provide near-real-time input for operational analytics can be selectively stored to help solve operational efficiency problems and support asset management.

Eyes and Ears

Data-analytic solutions are already well accepted, though not fully realized, within the utility as a key tool to improve reliability and avoid high-risk power outages. After the blackout of 2003 in the northeastern and midwestern United States, phasor measurement units (PMUs) were implemented to measure line condition at a rate as high as 30 times per second to avoid a similar widespread outage. The need for sensors is continuing to expand in response to the rapidly growing penetration of distributed energy resources (DERs) and the anticipated growth of plug-in electric vehicles (PEVs) at the connected home and office. As sensor technology becomes ubiquitous on the grid in homes and in commercial building management systems, the utility will have the opportunity for unprecedented visibility into the demand side of the equation, including the ability to shed load on discrete devices, enabling precise load shaping.

Utility stakeholders have been known to fondly call their sensors "eyes and ears." However, it is the ability to analyze the data with grid-specific models that is the brain for this ever-growing sensory system. This fact is not yet well understood, especially among device vendors who tend to concentrate on the functional characteristics of their units and how they can help grid operators build situational awareness. The growing availability of sensor data is advantageous to the operation of the grid in ways only limited by the ability to think resourcefully.

As an example of a novel use of sensor-data analytics, cellular base stations are significant consumers of energy, and the highly variable traffic load on the mobile networks constitutes a direct relationship between base-station traffic load and power consumption. Using sensors to understand this relationship can provide collaboration opportunities between telecommunications providers and utilities to identify energy-efficiency opportunities within the cellular access network. Without sensor data and analytics, these kinds of opportunities will continue to go unrealized. Data aggregation of sensor data with other forms of data will drive massive, unimaginable opportunity to make our business and living environments smarter, more sustainable, and efficient.

Smart grid digital-sensor technology has also expanded the capabilities for monitoring substation power-flow conditions and obtaining real-time reporting and analytics. However, the technology has been implemented quickly due to decreasing costs and, while many more data points are available, utilities have not been able to take advantage of installed monitoring functionality. Additionally, despite the substantial value that fault analytics hold, some types are not currently communicated back to the utility, resulting in missed opportunities to identify situations for rectification. Thus, some utilities find it difficult to create strong strategic business cases because the benefits expected from current implementations are not being recognized.

9.2.3 Control Devices

As in an organic system, once the grid can sense, it can respond. With the smart grid, communicating control devices allow the grid to responsively shed load during grid stress, maintain grid stability for managing complex DERs, and respond to unpredictable challenges to grid stability. The fully realized goal of smart grid control is called the "self-healing grid." This realization combines the visibility enabled by sensors, the flexibility provided by automated control devices, and the ability of embedded analytic software to automatically detect and isolate faults while quickly (within one to five minutes) reconfiguring the distribution network to minimize the impact of a grid disturbance. In one application of control devices, switches and reclosers on the distribution feeder will isolate the faulted section and allow service to be reestablished from alternate feeders or sources of generation. Control devices also facilitate the coordination of the grid in managing renewable resources, solar, and distributed generation.

The distribution area is especially improved by the deployment of control devices in the face of changing load dynamic on the distribution system. In heavily loaded systems, many of the distribution switches are controlled by either operators or by predetermined system settings. The control devices within the domain of advanced distribution automation, when combined with monitoring data, can be maximized by helping operators optimize the values in the systems necessary to provide volt-VAR support for improved decision-making.

Control devices are critical to the vision of smart grid automation, adjusting for power disturbances, providing the facility with remote repair, and delivering command and control from a centralized management system. Though this technology is key to the transformation of the modernized grid, outages and other problems that are detected, analyzed, and corrected in minutes must still be understood. Postprocessing analytics, which can be reconstructed from various sensors and intelligent devices on the grid, allow engineers to identify trends.

9.2.4 Intelligent Electronic Devices

Microprocessor based, intelligent electronic devices (IEDs) function as grid controllers among utilities and feature onboard capabilities that can receive data from sensors and other power equipment on the network; they can also issue control commands based on the received data. Typical uses of IEDs include tripping circuit breakers based on voltage, current, or frequency irregularities and the ability to serve as protective relaying devices, such as on load-tap changers, circuit breakers, capacitor bank switches, reclosers, and voltage regulators. The functions of IEDs within the grid infrastructure are varied and include protection, control, monitoring, and metering. The protection function covers a wide swath of grid-protection activities related to various faults, voltage, frequency, and thermal overload. The control features may be local or remote, and monitoring oversight can exist for various condition-monitoring and supervisory functions, such as circuits, switchgear monitoring, and event recording. IEDs also provide metering measurements for currents, voltages, frequency, active and reactive power, and harmonics.[2] Because IEDs are also able to communicate bidirectionally, it is possible to incorporate the data directly into the analytics life cycle.

IED data is especially crucial for root-cause and troubleshooting analysis because it provides extensive information every time there is a fault or an event. These recordings may include current and voltage waveform oscillography, the status of input and output contacts, the status of various system elements, and other settings. Overall, the data characteristics of IEDs create excellent observability and analysis potential due to their abundant and redundant measurements, resulting in improved fault analytics and visualization of event data.

9.2.5 Distributed Energy Resources

The rising penetration of DERs—including renewables, microgrids, EV networks, and storage—on the grid increases the possibility of disturbance, from voltage-control issues to intermittency in the energy supply. The application of smart grid data analytics to the management of renewables is one of the most powerful use cases of advanced modeling to control and monitor DERs to ensure reliability. To successfully monitor the grid under the conditions created by DERs, the utility must have real-time information, excellent situational intelligence, an understanding of prevailing weather conditions, and the ability to integrate that data to make informed and rapid decisions to manage frequency control, power quality, and other operational parameters.

In fact, without analytics to manage DERs, the rapidly increasing interconnection of these resources into the macrogrid could lead to unexpected events of large magnitude and consequences. In the rush to green the grid and increase regulation over dirty baseload generation, if the utility cannot successfully cope with the effects of intermittent renewables in the generation mix and within the delivery network, no amount of hindsight can account for risks to provide a balanced grid. DER integration, especially for renewables, requires a deeper and more immediate context than postprocessing for fault or root-cause analysis. Successful renewables integration requires the utility to account for wind, cloud cover, and other environment variables on the generation sources themselves. These factors can change instantaneously, and in order to align demand with capacity, the utility must be able to have a high level of confidence in its forecasts for projected mixes of energy sources. Forecasting is not

[2] Raheel Muzzammel (n.d.), *Intelligent Electronic Devices*. Academia.edu. Retrieved from http://www.academia.edu/1739791/Intelligent_Electronic_Devices.

perfect, and even overcapacity conditions can lead to brownouts by pushing excess power upstream.

In addition to real-time weather data, analytic models for managing DERs include power-line-sensor data, currents on primary and secondary feeders, voltage, and currents on the primary and secondary sides of the transformer, as well as other transformer parameters that increase predictability for safe and reliable grid operation. Apart from assisting operators in making faster, better decisions, DER intelligence can help prospect appropriate sites for new generation, optimize the generation and transmission of assets, and increase the confidence level in forecasting capabilities over time.

9.2.6 Consumer Devices

As alluded to earlier in the chapter, the proliferation of devices across the grid has transgressed the traditional demarcation of the meter into the customer's structure, effectively expanding the breadth of the grid directly into the home, commercial building, campus, and industrial enterprise. This explosion of Internet Protocol (IP)-addressed devices that are being sewn into clothes, watches, stereos, building controls, and smart appliances is called the Internet of Things (IoT), introduced in Chapter 3. From an analytics perspective, this equates to a massive volume of data that describes a building and the behavior of those within the building as they interact with energy-consuming devices. Some of these devices are designed specifically to provide opportunities to save money and energy by reducing energy consumption and shifting demand. However, any sensor that can meter consumption can be analyzed, modeled, and leveraged to provide a variety of benefits to the utility.

Collecting and modeling data off of demand-side devices with monitoring capability is the leverage point for utilities that are seeking to build trust with consumers, as well as mitigate the risk to their core business model by offering new products and services. For example, energy models from interval consumption data over time can indicate when an appliance, like a refrigerator, is in a state of disrepair and will be experiencing imminent failure. There is tremendous value in that kind of information, and those who live in broadband-enabled households are expressing an increasing level of interest in smart washers, dryers, and air conditioners.[3] Connected devices bring control and convenience to the consumer; for the utility, the low-hanging fruit is the ability to signal smart appliances to shed load in real time and to verify that load.

---

[3] Parks Associates (2012), *Energy Management Devices: Engaging Consumers.* Retrieved from http://www.parksassociates.com/services/energy-devices.

It is not clear with the emergence of the connected home what role the utility will directly play in the provision of added-value energy services; it's also unclear whether those IoT devices will be under utility, consumer, or third-party control. It is certain, however, that the market is being rushed by telecom and cable providers seeking to offer home controls, security, and even energy management. Up until now, the utility has focused primarily on demand-response applications to reduce consumption or shave peak. The pilots have been extensive, but the rollouts have been few. Utilities are learning to collaborate with connected-device experts who understand marketing and already have a presence in the consumer's home.

Analytics based on source data from the demand side can help solve myriad utility business problems, including managing the interconnection of micro- and nanogrids, performance monitoring, highly refined appliance-level demand response, dynamic-pricing programs, and PEV management. The analytics derived from consumer data will help utilities expand their market potential and find new ways to drive revenue. Additionally, combining such information with the plethora of data sources available about consumers—including demographic data, behavioral web and social data, allocation data, and financial records—will provide the foundation for the next level of insights for predictive utility applications.

9.2.7 Historical Data

No one could reasonably argue that there is a growing need to retain data and improve access to it. However, many utility stakeholders are concerned that keeping all the data that flows into the enterprise is simply cost-prohibitive from both the physical storage and management aspect. And indeed, stashing it in offline archives limits the ability to extract any value from it, especially with analytic workflows. This problem also has two difficult relatives: compliance and privacy. Compliance is related to mandated reporting requirements, while privacy is related to policy and governance for how long personally identifying information is stored, in what form it's kept, and under what conditions. Aggregated data that has been scrubbed and anonymized is not subject to most privacy regulations. However, in the context of solving utility business challenges related to how and when customers use energy—particularly in support of improving energy efficiency and demand response—the information becomes unserviceable. In the quest to overcome privacy concerns, utilities are interpreting laws and rules, often resulting in reluctant and conservative positions toward data.

The usefulness of historical data relates directly to how it's collected, organized, and stored. Because it tends to presuppose the way the data will be queried, overly normalized data restricts how it can be used in analytical models, limiting future insights that might be gained by novel views and uses, especially predictive and prescriptive applications. In some jurisdictions, there is serious discussion of collaborative efforts that would create an "energy data center" that provides aggregated and anonymized data to the public. The US state of California has recently proposed such a project as a way to ease access to customer energy-usage information for consumers.[4]

This is a very interesting proposition for the utility as well, as it would likely lead to a spate of new and interesting energy applications for consumers. It would also lead to the creation of new market opportunities for energy-saving products and energy market research. Moreover, such an energy data center would relieve the utility from having to manage customer-consent processes to share data, and it would provide insight into an industry-standard methodology for aggregating and securing consumer information. For the utility that is seriously considering future strategic moves into enhanced energy products and services, it can reposition itself as a barrier against sharing data, and it can find explicit ways to benefit from the research and wisdom that are unleashed when information is made public.

9.2.8 Third-Party Data

The discussion of historical data naturally leads to the issues and concerns related to the use of third-party data in utility analytics programs. Much of the discussion about third-party data concerns sharing customer information that has been collected by the utility, especially billing and smart meter consumption data. But, in the sense of a data source—especially for predictive analytics—fusing third-party data such as weather, new-customer demographic information, premise data, social graph data, financial records, mobile data, and GIS data with internal sources of information can help treat several utility business issues, including:

- **Customer microsegmentation.** Segmenting customers based on patterns of data.
- **Demand forecasting.** Improving predictability for more-optimal planning.
- **Fraud identification.** Providing a more expansive view into revenue leaks.
- **Program optimization.** Determining which customers the utility should be targeting for better program uptake and outcomes.

---

[4] Audrey Lee and Marzia Zafar (2012). *Energy Data Center*. California Public Utilities Commission. Retrieved from http://www.cpuc.ca.gov/NR/rdonlyres/8B005D2C-9698-4F16-BB2B-D07E707DA676/0/EnergyDataCenterFinal.pdf.

There are already analytics companies that are targeting utility problems for the application of predictive data science to meet these needs. They are literally aggregating thousands of data points from hundreds of sources. Because of the fragmented nature of the utility industry, where utilities do not share data for operational and business benefit, data aggregators will step in. These aggregators will coordinate the volumes of big data and use their proprietary models to provide insights from the data. As utilities assess vendors that may be able to help bring third-party data sources to bear, it is worth considering the value of sharing anonymous data from many utilities to improve analytic outcomes. This won't always be useful, as utilities have uniqueness among their customers, geographies, and technologies. However, in many cases, sharing data could greatly accelerate understanding about how to incentivize customers to participate in dynamic-pricing and demand-response programs, as well as help utilities improve their own metrics by understanding how they are performing compared to their peers.

9.3 Working with a Variety of Data Sources

9.3.1 Data Fusion

Data fusion is a core capability for working with many data sets within the discipline of predictive data science. As simply described in Figure 9.1, data-fusion technologies and approaches merge disparate data sets and manage conflict resolution among structured, unstructured, and streaming data sources

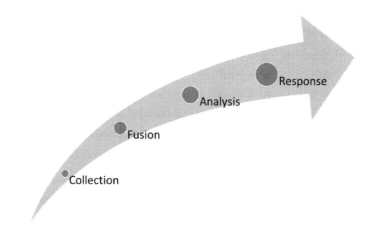

Figure 9.1   The Role of Data Fusion Is Aggregating Complex Data Sources.

so that analytic models and algorithms can be rationally applied for response and action. Despite its synthetic nature, data fusion can be more informative than data sets considered in isolation; done improperly, it can be distracting and misleading. The process of fusion may either be low, intermediate, or high, depending on the stage at which the work is done. For example, low-level data fusion might combine several sources of raw data to generate an entirely new raw data set that is used for analytic processing. High-level data fusion considers the data at the object level, and it will fuse information at the level of relationship among those objects. Utilities are already familiar with high-level fusion; the power-plant control room is a functioning fusion center, since it manages relationships among sensor data, humans' behavioral data, and physical objects on the network that may be impacting the grid in real time.

Thus, data fusion aggregates varied data types, including structured, semistructured, and unstructured data sources, into an aggregated form that can be modeled. Structured data is very common in the operational context because it may be either machine-generated messages produced without human intervention or data created by a human through an interaction with a computer application. In either case, the data is usually well understood contextually and at the record level. Structured data, such as information produced by sensors, financial systems, applications, and clickstream data, is often astronomical in size but tends to consist of similar, consistent, and expected information. Because of these characteristics, structured data is more easily stored and queried within the context of traditional relational databases. Unstructured data also can come from machines and humans, but it consists of unexpected information, such as satellite imagery, videos, social media data, and all the text within the utility enterprise that comprises documents, log files, and e-mails. The uses for and the ability to readily process unstructured data are rapidly improving. However, some of the most powerful use cases for unstructured data come from the ability to fuse it with structured sources of information to extract highly relevant insights at a granular level. Unlike unstructured data that is unpredictable, semistructured data sources are self-describing. And unlike structured data that has fixed records, semistructured data is schemaless and nonconforming. It sits somewhere in the middle, requiring a different approach to processing. Markup languages such as Extensible Markup Language (XML) and electronic data interchange (EDI) are examples of semistructured data.

High-level data-fusion computational models are often described in one of three ways: physics based, data based (knowledge lean), or knowledge based (knowledge rich). Physics-based systems rely on both linear and nonlinear equations to specify behavior in the model (such as Kalman filtering or Sequential Monte Carlo methods), data-based fusion that relies on input/output variables to extract system-behavior models (such as machine learning), and knowledge-based models are founded on an ontological understanding of system behavior (like fuzzy logic).[5] As in the analytical systems that leverage the information processed with various fusion techniques, most fusion systems rely on a combination of the fusion-modeling systems to filter and associate data.

Data fusion is a key process in utility data-integration efforts. Although it is typically thought of as a way to combine disparate data, it can also be viewed as a way to reduce (or even replace) the volumes of data—while actually improving confidence. Data fusion is widely used in advanced data-integration projects, including GISs, business intelligence, wireless sensor networks, and performance management, and is a key component in preparing raw and historical data sources for advanced analytic applications.

Next Steps for Utilities
• Develop use cases. Don't invest in big data technologies until use cases are developed for how the data is going to be used.
• Determine whether the internal data being collected is enough to solve the use cases. If it's not, what data is necessary and how can it be obtained?
• Understand the required data completely, what data sources are required to address business problems, and which types of analytics are appropriate for the data sets.
• Don't be tempted to underestimate the value and costs associated with sourcing data: It must be collected, understood, structured, and fused.
• Scope the project in business terms based on data availability, and plan from there.

[5] Subrata Kumar Das (2008), *High-Level Data Fusion* (p. xvi). Artech House, Norwood, Massachusetts. Retrieved from http://books.google.com/books?id=iTb3e9efuoMC&pgis=1.

Chapter Ten

Big Data Integration, Frameworks, and Databases

Scientist examines the slosh tank apparatus in the NASA 10x10 wind tunnel shop. (*Source:* NASA[1])

10.1 Chapter Goal

In traditional database integration and storage efforts, there has been a clear demarcation between data storage and data processing. With the advancements

---

[1] Image retrieved from the public domain at http://d3.static.dvidshub.net/media/thumbs/photos/1210/709412/288x360_q75.jpg.

in efficiency and performance technologies, this line is beginning to blur. In this chapter, we discuss the elements of big data infrastructure from the perspective of existing approaches, their difficulties in adapting to the needs of high-volume and varied data types, and the benefits of distributed approaches that are more cost-effective. The open-source big data technologies—Hadoop, the Hadoop Distributed File System, and MapReduce—are described, as well as other database technologies that are beneficial within the utility ecosystem. We'll also address the fundamentals of different database concepts, their defining characteristics, and the best use of each.

10.2 This Is Going to Cost

Many utility stakeholders believe that big data requires a big check. Learning to optimize big data across the enterprise is one approach to controlling these costs. Fragmented department-level projects are never the most efficient, cost-effective way to formulate an enterprisewide analytics strategy because they lead to unchecked technology iteration without a cohesive and comprehensive vision. However, having an understanding of the key pieces and parts of this vision will help streamline the development of an optimal strategy.

In a complex ecosystem like the utility—which is process driven and outfitted with large, expensive, and fast computers—terms like Hadoop and MapReduce can seem like flashy jargon that has no place in the real-world domain of power delivery. But if the organization is seriously interested in data analytics, this argot becomes very important. At the same time, these names aren't really critical to understanding, at a detailed level, how these technologies function; what is important is understanding their benefits. And to understand the benefits of big data management systems, there are two important topics of interest: data storage and data processing, which we will discuss in the next two chapters. Unlike traditional data management, in the world of big data, these often happen in the same system.

It is certainly not a new phenomenon in the energy industry to interrogate a large database to gain business insight. But these exercises have been almost exclusively performed in a data warehouse or a high-performance computing (HPC) system on structured data with high latencies, in batches, overnight, or even sometimes over the course of several weeks. Big data deals with rapid access to many different forms of data, including unstructured and semistructured data, and the value proposition of big data analytics is found in the immediate response times to time-sensitive analytical queries.

10.3 Storage Modalities

The core requirements for big data storage are capacity, scale, and high-performance input/output (I/O) operations per second, or IOPS (pronounced "eye-ops"). None of these requirements is straightforward. For example, IOPS performance is highly dependent on system configuration, operating system, and innumerable other factors. And when very fast response times are required with massive volumes of data, traditional scaling approaches are no longer enough. There are several approaches to meeting big data requirements.

10.3.1 Hyperscale

One of the answers to big data applications is hyperscale. These environments are built using inexpensive servers and storage that are connected in a single system. Storage units in the environment are directly linked to the servers in an approach known as direct-attached storage (DAS). Because DAS does not have to traverse the network in order to read and write data, it's used in high-performance environments. Redundancy is provided at the storage level, so if any device fails, immediate failover to a mirror unit occurs. For even faster response times, flash storage may be implemented in addition to fast disks. A hyperscale environment isn't always necessary to mine customer metrics or perform simple business functions; however, grid operations or high-intensity computational finance functions easily benefit from such a specialized environment, particularly as the volumes of data grow.[2]

10.3.2 Network-Attached Storage

Technical operations that shy away from hyperscale may opt for shared storage access with network-attached storage (NAS) or a clustered NAS system. This approach can adequately meet the big data storage demands of capacity and low latency, depending on the speed of growth of the data and access demands. Clusters of NAS boxes can be configured in a gridlike collection of nodes that aggregate processing power in a parallel configuration. At its simplest, NAS is a file-level computer data storage solution running a stripped-down operating

[2] Antony Adshead (2013), "Big Data Storage: Defining Big Data and the Type of Storage It Needs," *Computerweekly.com*. Retrieved from http://www.computerweekly. com/podcast/Big-data-storage-Defining-big-data-and-the-type-of-storage-it-needs.

system that's connected to the network for data access. NAS devices are highly specialized and manufactured specifically as a computer appliance that's optimized for storing and serving files. Yet, traditional NAS configurations will not scale to massive levels, and they have independent file systems that cannot be searched as a single unit. In a cluster configuration, NAS systems can have petabyte-level file system capability that does not degrade because the system can easily grow as processing nodes are added. Traditional NAS solutions are still quite popular, and prices have been plummeting. They are also easy for IT staff to manage and configure with a NAS management tool.

10.3.3 Object Storage

The other form of storage that is used in big data environments is object storage. Instead of files, objects contain data, but they are organized in a sort of hierarchy. Managed by a system of metadata that describes each of them, the objects maintain unique identifiers, which simplify data storage and access. Think about the last time you checked your coat in a restaurant. The coat-check clerk took your coat and handed you a ticket. You didn't know (or likely care) where your coat was placed or if it was moved several times over the course of the evening. What mattered was when you went back to retrieve your coat and you handed the clerk your ticket, he expeditiously and efficiently returned your item. That's object storage. Simply, the system gets around the issues of becoming unmanageable under the reality of growing data by relying on an index of metadata to handle a large number of files. Object storage technology is younger than NAS technology, and it can be scaled to massive volumes in a reliable manner. However, it has drawbacks, mostly related to slower throughput and the time to establish data consistency. Object stores may be an excellent fit for the money, particularly for data that isn't rapidly changing, such as media files and archives.

10.4 Data Integration

Very few disciplines are still developing after 30 years, yet the practice and theory of data integration just doesn't stand still; it is still evolving because database implementations change. First there were flat files; then there were hierarchies, relational databases, object-oriented databases, Extensible Markup Language (XML); and again, flat files (but this time with different management strategies). In fact, in the world of big data, the term "database" is not even entirely accurate anymore because the data is now stored under the auspices of a "framework" that includes both a way to organize data in a file system (like a database) and processing capabilities.

This shift in focus is driven by two factors: the requirement that collection happen faster and on more data sets and the need that it be optimized for rapid analysis. But, despite the fact that data integration is the choke point in unified systems, many big data projects will falter at the level of data integration simply because of a lack of functional requirements—there is just no clear sense of the desired behavior of the system when it is being built. By definition, data integration is the process of combining various data sources and types to provide a unified view of the data within the storage system. In the scope of big data, integration is a problem that becomes more difficult as the variety and volume of data explode. In fact, it is data integration that ensures that the technology solution for the big data problem actually supports the business requirements. The process must be well thought out and trusted.

After the utility has established which data store is appropriate for the data analytics project in question, it is time to begin assessing data integration approaches and processes. Strangely, the hype around big data integration is almost as large as big data itself, and just as difficult to pick through. The best data integration process for an enterprise is that which meaningfully combines technical and business drivers into a cohesive process. Several key characteristics define a successful big data integration solution, including a method for data discovery, cleansing, transformation, and movement of the data from its source to the data store.

It should be clear now why many companies have been distressed to learn that building your own big data storage and integration environment can be an expensive mess. If big data is going to be cost-effective and sustainable, a repeatable, coherent approach is needed.

10.5 The Costs of Low-Risk Approaches

Lead me not into temptation; I can find the way myself.
– Rita Mae Brown

While data integration has been a changing process for decades, it has always had a consistent goal of extracting data from multiple sources, transforming it, and consolidating it to create a unified and consistent version of truth. Workflows were developed and graphical and easy-to-use tools helped design integration routines on constrained amounts of data that didn't require custom, nonstandard, and expensive-to-maintain software. With big data, the problem with this approach is that there is a constantly growing amount of data from a wider variety of sources. Therefore, data integration is no longer a problem of finding and deploying function-rich tools; it's a problem of efficiency and performance at an unprecedented scale. Many of the data integration best practices, while applicable in many ways, simply do not meet the big data requirements.

In moving quickly to establish big data analytics-integration processes, many utilities will be tempted to scale in ways that are costly and do not deliver the necessary benefits. For example, some organizations have attempted to deploy a staging area, or landing zone, that is designed to improve the efficiency of existing extract, transform, load (ETL) processes. Staging areas may consist of a relational database; collections of XML files; or some type of file system organization where precalculations, data cleansing, and other forms of consolidation may take place. Depending on the implementation, this approach can be quite expensive just in database infrastructure and maintenance costs alone, and it can unnecessarily increase the complexity of the data integration system overall.

The temptation to control the budget with a rush to recognize short-term results leads to suboptimal approaches that are not sustainable and inevitably cost more. This is a losing battle. Consider the fact that even if your incremental effort to drive performance of an existing system results in a 50 percent gain, if it used to take a week to run a process, the utility may still be running into two- or three-day cycles to work through a job. In the world of ever-flowing data, by the time the process is complete, the data may very well be obsolete.

Utilities that have begun serious big data efforts are beginning to reckon with spiraling costs. Why? The authors at TDWI (www.tdwi.org) describe the phenomenon of the tail-chasing cycle that happens as organizations attempt to buy their way out of their efficiency and performance problems by adding more hardware, software licenses, power and cooling infrastructure, and staff:

> After a while, people realize they cannot hardware their way out of this problem . . . At this point, it's like taking a huge step back in time—data lineage is lost, the database is overloaded, and costs and complexity rise to the roof.[3]

10.6 Let the Data Flow

In the scope of acquiring and coalescing the preponderance of data that makes up the world of big data in the utility, new approaches to data integration are

---

[3] J. Lopez (2012), "Big Data Integration," *TDWI*,Syncsort E-Book. Retrieved from http://tdwi.org/research/2013/12/best-practices-report-predictive-analytics-for-business-advantage/asset.aspx?tc=assetpg.

needed. One way to think about it is to focus on the flow of data rather than the old-school integration of data, where data unification and loading are primary squeeze points in the process. This moves the organization away from a mindset centered on data integration as a staged process and toward frameworks and environments that consolidate the principles of ETL into a single solution. In this way, data mapping, data loading, and accessing enterprisewide data across mixed application environments are truly efficient. However, as mentioned previously, the best way to integrate the wide variety and high volume of information that the utility requires for a comprehensive big data analytics program depends on the framework that's in use. Many utilities are looking for these capabilities within a single solution that can provide unified, high-quality, trusted information. And as the need to keep data flowing increases, the process of integration is merging into the big data environment much more than data warehouses, master data management systems, and custom applications ever could.

The utility stakeholder embarking on a program of big data is exposed to a new and expansive patois of bizarre and unfamiliar terms and all their strange friends. Many of these terms represent powerful technologies, but like the jargon before it (for example, "mashups," "brain dump," and "crowdsourcing"), sooner or later we adapt and never think about these strange descriptors again. While this list barely scratches the surface of the extensive glossary of big data terminology (some have called it a bestiary), following are some of the key terms and concepts that need to be understood to further the discussion on big data frameworks.

10.6.1 Hadoop

The big data revolution has been brought to you compliments of Apache Hadoop (hadoop.apache.org). An open-source framework for storing, processing, and subsequently analyzing data at a low cost and on a massive scale, Hadoop enables the storage of an enormous quantity of data across a distributed cluster of servers, and allows users to run analysis applications on those clusters.

Hadoop emerged in 2006 from a Yahoo!-funded project created by Doug Cutting, who named this fault-tolerant, scalable, distributed computing system after his son's stuffed yellow elephant (meaning, it stands for nothing). Hadoop can scale across literally thousands of commodity services to provide resilient storage and processing of large data sets in a distributed environment. The technology transformed the world of big data by changing the economics and dynamics of large-scale computing. How? Imagine how much it would cost your organization to buy a 1,000-central-processing-unit (CPU) machine versus 1,000 single-CPU machines tied together in a cluster. Now, just for a

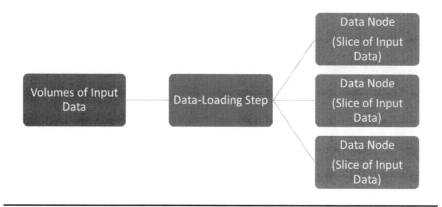

Figure 10.1   Data Loading Using the Hadoop Approach.

nauseating moment, assume that your awesome and expensive machine with 1,000 CPUs just failed. Enough said.

Conceptually, Hadoop is quite straightforward. As described in Figure 10.1, when data is loaded in, it is distributed to all the nodes of the cluster by splitting large data files into smaller chunks, which are managed by different nodes. Additionally, each chunk is replicated across several machines, so that if a single machine fails, the data is still available. With Hadoop, the data is record oriented. This means that as individual input files are broken up and stored, when the records are processed, they are running on a subset of the data. Thus, application computations can be scheduled according to the data that is closest to the processing node, driving down latency by moving data around the network and keeping computation close to the data, as opposed to shipping data to a specific device for computation.

It is through this notion of data locality that Hadoop achieves its performance and scalability. As you may have guessed, based on our previous discussion about storage modalities, DAS is an assumption in the Hadoop environment. Why introduce the extraneous processing required by NAS and negate the benefits of data locality? Don't yet count out the role of NAS for secondary storage or even for primary storage implementation. Either way, the principle of replicated blocks of data across multiple machines drives each individual compute process to be isolated from one another. The way this disparate data is processed is through the use of a model called MapReduce.

10.6.2 MapReduce

MapReduce is a model for processing large data sets using a parallel, distributed algorithm on a cluster of computing devices. Most people think of MapReduce as the way in which data is processed in a distributed environment. This is true; however, MapReduce can also be successfully used to manage heterogeneous data sources, especially when complex computations are required on large amounts of data that need to be integrated. There are many implementations of MapReduce, and Hadoop is one of them. Basically, the "Map" part of the model is responsible for splitting up the problem, and the "Reduce" part puts everything back together to compile a single answer. It works like this: "Map" splits the problem into smaller parts using a series of key-value pairs (so it can find them later), sends those parts to different machines within the cluster, and then runs all the pieces in parallel. "Reduce" steps in again and finds all the values that have the same key, and then combines them into a single value.

Here's an example: A nonprofit research company wants to count the number of Fortune 500 companies that have women as corporate officers. Ultimately, it needs to know who all the women are in the organizations who hold these top-earning positions and then aggregate that information. There's an obvious methodology for gathering this research: The team has many researchers who each are given a list of companies and instructed to collect the required data and return it at the end of the day. These researchers are performing what Hadoop has named "mapper tasks," where each company they are researching (we would call this a key) may have several women in these roles (we would call these values). Then, imagine that when all the researchers send in their collected data to the project manager (Hadoop would call the manager the reducer), he creates a spreadsheet of all the information. Of course, analysts will use this information to draw all kinds of conclusions, but the point is that the project manager and the team of researchers executed a classic MapReduce algorithm.

Figure 10.2 demonstrates this process. To be sure, MapReduce isn't exactly rocket science, but when faced with millions (sometimes billions) of rows of data that might not be so neatly structured, the value of performing these operations in a massively scaled environment is profound.

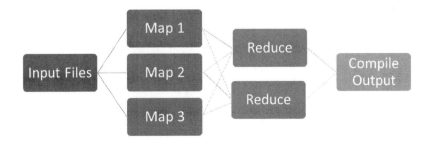

Figure 10.2 The MapReduce Algorithm.

10.6.3 Hadoop Distributed File System

While MapReduce is the programming model used within Hadoop, the Hadoop Distributed File System (HDFS) is, as its name implies, its own file system. However, in the spirit of open source, several file systems are supported out of the box by Hadoop, including Amazon S3, CloudStore, File Transfer Protocol (FTP), read-only Hypertext Transfer Protocol (HTTP) file systems, and HTTP Secure (HTTPS) file systems. In fact, Hadoop can work with any distributed file system that can be mounted using the file://URL (uniform resource locator), but not without a performance cost, as it is the Hadoop-specific file system bridges that maintain the advantage of data locality. Remember, Hadoop performance relies on knowing which servers are closest to the data.

It is actually a function of the HDFS to split data into chunks to be managed by different nodes within the cluster. And, although it's not required, the data is redundant; each chunk is replicated into smaller pieces across the multiple data sources to better benefit from data locality Specifically, HDFS is designed to reliably store very large amounts of data and provide fast access for reading and computation, at scale.[4] HDFS is best suited for streaming read performance, unlike a database that allows files to be updated, modifications are not supported (although appends are), and random seek times are not optimal, unless extensions such as HBase (a purpose-built indexer that resides on top of HDFS), which provides fast record lookups, are employed.

10.6.4 How Does This Help Utilities?

Electric utilities that embrace open-source projects are among those that have had early success with Hadoop. As recently as 2009, the US's largest public provider, Tennessee Valley Authority (TVA), serving more than 9 million customers in seven states, has been using Hadoop to collect data from phasor measurement units (PMUs). Data is collected from the field from nearly 1,000 PMUs at intervals of 30 times per second. The information is captured as time-series data and sent to Hadoop for processing, where it is run en masse. TVA selected Hadoop not just because of its ability to reliably process large volumes of data and store it, but more importantly, because of its ability to scale out cost-effectively.[5] Other utility organizations may find that although

---

[4] Hadoop Tutorial (n.d.), *Yahoo! Developer Network*. Retrieved from http://developer.yahoo.com/hadoop/tutorial/module1.html.
[5] Dave Rosenberg (2009), "Open-Source Hadoop Powers Tennessee Smart Grid," *CNET News*. Retrieved from http://news.cnet.com/8301-13846_3-10393259-62.html.

Hadoop isn't perfectly suited to their overall ecosystem, it is certainly appropriate to accommodate complex data-integration problems by splitting up tasks before the data is ultimately pushed into a data warehouse. This approach would likely encompass a complete integration redesign to compress processing times, as opposed to extending the ETL runway to buy time as traditional staging areas often do.

There is an important caveat here: Do not use Hadoop just because your data is too big to load in an Excel spreadsheet or because it's free. Despite the tool's big-data-darling status, distributed processing is not always the right answer. Hadoop is evolving quickly with new functionality and facilities, so it pays to remember that even though the cost per unit of data may be lower with Hadoop than a relational database, clusters of servers and specialized employees who have advanced programming and data management skills may not really be that cheap for every situation. Traditional tools, specialized capabilities such as in-memory databases, or even relational databases may even be better suited to the needs of the business. And time-series database servers, spatial databases, or geographic information system (GIS) databases may be more appropriate solutions for the problem at hand.

Understanding the business problems first, before assessing the benefits of various solutions, increases the likelihood of a successful implementation when it comes to choosing the right technologies for the job. Besides, make sure you are committed to Hadoop before investing in engineers to develop a solid understanding of the mysterious Hadoop enablers: Pig, Hive, Sqoop, and Oozie.

10.7 Other Big Data Databases

Discussions about the shortcomings of relational databases for big data are commonplace when the topic of big data storage comes up. It's fairly straightforward as to why this is—these systems are designed to manage structured data. In fact, structured data is often called relational data and is well suited to an organization that needs specific parcels of information to be organized in rows and columns. Relational database management systems (RDBMSs) typically incur expensive overhead requirements during processing cycles because the data is searched by the actual content that is stored within any particular field in the database, such as a customer's ZIP code or other discrete information. However, big data analytics leverage both unstructured and structured data, including free text, images, objects, and other types of raw information. Thus, if scale and flexibility are required, then relying solely on RDBMS technology will be a less-than-ideal solution. There are many other options, even if the use of an RDBMS is part of the overall approach.

The following is a catalogue of some of the most prominent big data databases used by utilities outside of the popular Hadoop technology. While certainly not a comprehensive directory of all useful technologies in the realm of big data analytics, the major approaches that can be valuable within the utility ecosystem are outlined. Each technology is described for its approach, defining characteristics, and the best use of each.

10.7.1 NoSQL

As alluded to, it's not possible to write off relational technology because it's still important to the enterprise, and it's simply a fundamental business requirement to be able to create relationships among data. This does not mean that nonrelational database technologies are all hype. They have actual benefits within the big data analytics environment. We've already discussed Hadoop and described it as a distributed computing ecosystem; Not Only SQL (NoSQL [Structured Query Language]) is a broad class of database management systems that may actually be deployed within the Hadoop milieu. Unlike relational systems, NoSQL databases do not use a fixed schema for data organization, and they run well in replicated situations and distributed situations, lending themselves to the scale required by big data applications. There a several variants of NoSQL databases usually categorized by the appropriate data model. These include:

- Key-value stores that support very fast simple retrieval and appending operations
- Document databases, where the value in the key-value pair is a complex data type, known as a document
- Graph stores that contain network data such as social data connections
- Wide-column stores that support queries against large data sets that are stored in columnar format (as opposed to the RDBMS rows)

Unlike relational databases, most NoSQL implementations do not guarantee that transactions process reliably. Instead, a paradigm called eventual consistency is employed. As implied in the name "NoSQL," these systems do not use SQL for interrogation; rather, they may use lower-level languages or application programming interfaces (APIs). For big data implementations, the benefits of NoSQL databases include the ability to scale out—meaning that commodity servers (or clouds) are used to add capacity—and the ability to dynamically manipulate a schema, including adding new information types to records on the fly.

10.7.2 In-Memory or Main Memory Databases

Another strategy being used for faster processing of large data volumes is the in-memory database (IMDB) or main memory database (MMDB). The IMDB relies on the main memory of the computer as opposed to disk storage for data storage. These databases are especially useful for vertical applications, delivering benefits that are derived from simple, internally optimized algorithms that execute fewer CPU instructions. IMDBs leverage volatile memory and are useful where response time is critical, but they lose all information in the event of a power loss. However, with the advent of Non-Volatile Dual In-line Memory Modules (NVDIMMs), this data loss is increasingly less of an issue, since these modules allow IMDBs to achieve the consistency and durability of a traditional RDBMS. Because these systems increase processing speeds and data handling by eliminating the mechanical activities of disk I/O, these databases are especially appropriate to meet the performance needs of the devices found within the smart grid and connected homes.

IMDBs come in many variants, including SQL relational, NoSQL, and distributed. Utilities may already be familiar with the IMDB database, as the Polyhedra IMDB is a common storage solution for supervisory control and data acquisition (SCADA) and embedded systems. Among the benefits of IMDBs is the elimination of seek time during queries, which has been measured to be between 10 and 100 times faster than conventional databases.[6] For high-availability implementations, IMDBs tend to be used in conjunction with other mechanisms to provide failover and data replication. The databases support the use of both structured and unstructured data and are thus advantageous to in-memory analytics, where scenarios or complex calculations can be run extremely quickly. This response time is especially useful for analytic visualizations that support real-time modeling and data exploration that have previously been constrained by slower computation times. IMDBs are best considered for targeted solutions to solve specific business problems that involve a high volume of data with a wide variety of data types.

10.7.3 Object-Oriented Database Management Systems

Object-oriented database management systems (OODBMSs) manage information as objects in order to augment the use of object-oriented programming

---

[6] Chris Preimesberger (2013), "In-Memory Databases Driving Big Data Efficiency: 10 Reasons Why," *eWeek*. Retrieved from http://www.eweek.com/database/slideshows/in-memory-databases-driving-big-data-efficiency-10-reasons-why.

paradigms with corresponding database technology. An OODBMS will typically allow object-oriented applications to store data as objects, and to replicate or modify existing objects directly in the database. Not normally considered when conceiving of big data management technologies, OODBMSs can support graph-structured data types and are well matched to the management of complex data types. They are useful in the utility context because they support engineering functions, including spatial applications. Object data is accessed directly, without the mapping required by RDBMS, and in applications where the stored objects have many-to-many relationships with other objects in the system. In-memory and NoSQL implementations of object-oriented databases also have been developed, but the role of OODBMSs in big data and analytics is not yet well defined.

10.7.4 Time Series Database Servers

Time series database servers (TSDSs) are systems specifically designed to handle time-series data, which are successive data points that are measured at points in time, typically spaced at uniform time intervals. Because time-series data is used in signal processing, moment-to-moment weather forecasting, and control and communications engineering, it's especially relevant to grid operations. For example, load profiles are a time series of energy-consumption values. Time-series data is not naturally relational, nor is it always well suited to flat files if the data volumes exceed the capabilities of the underlying system. A TSDS is purpose-built to optimize the handling of the streaming characteristics of time-series data. These systems can be built on top of existing technologies, such as Apache HBase, and are tuned to meet analytical requirements of statistical operations. TSDSs are the appropriate underpinnings for high-performance historical analysis and have already demonstrated their usefulness in the utility sphere, where they have been used to process the data from millions of smart meters and smart grid devices to calculate load in minutes rather than hours. Meter data management is a prime use case for the appropriateness of TSDSs because the systems reduce storage and system costs while providing optimal processing with linear scalability.

10.7.5 Spatial and GIS Databases

Spatial and GIS databases are optimized to specifically store and process data that describes objects that exist in geometric space. These are represented as points, polygons, and lines. Some spatial databases are even capable of handling three-dimensional (3D) data, including surfaces and topological coverage. Spatial and GIS databases may be implemented as overlays on existing database systems and have been useful to the utility for many years. GIS data, in one form or another, has been leveraged for decades to manage the outside plant. Within the context of the smart grid, this type of database has become even more significant. Within the context of analytics, databases that can account for grid assets are an absolute necessity in driving better decision-making opportunities.

Using optimized spatial indexes that support functions related to measurements, intersecting features, and the construction of new geometries, spatial databases are usually expected to be OpenGIS compliant, though not all are. This means they reflect the standards set by the international Open Geospatial Consortium (OGC). These standards have been implemented on NoSQL, relational databases, graph databases, and purpose-built systems.

10.8 The Curse of Abundance

Certainly, there is an abundance of big data integration and database technology. And it's difficult to make the issues less overwhelming to those not immersed in the field. However, in this case, the bad news is also the good news for utility strategists. If nothing else, it is patently clear that there is not one single solution for all the data analysis projects in the utility, but it's important to exercise informed choice.

Planning Questions
1. What are the most important features that will improve productivity for the IT team, application developers, data scientists, and analysts?
2. What are the major constraints in the desired system in terms of latency and access to the data insofar as it affects the objectives of the project?
3. What existing enterprise data sources must be integrated into the big data platform, such as relational databases and various enterprise applications?
4. What are the requirements related to data consistency, durability, and availability?
5. What needs to be processed, where does it come from, and who needs to use it?

Chapter Eleven

Extracting Value

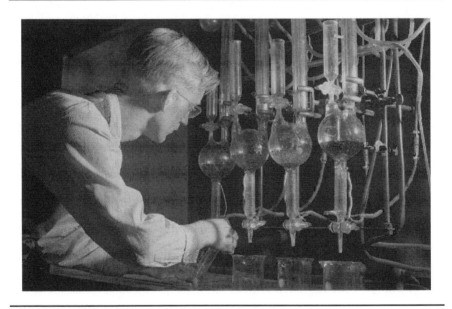

Tetraethyl lead extraction apparatus used by NASA scientist. (*Source:* NASA[1])

11.1 Chapter Goal

Successfully extracting value from utility data is dependent on effective processing techniques and the ability to find the correct algorithms for the right

[1] Image retrieved from the public domain at http://d3.static.dvidshub.net/media/thumbs/photos/1302/860344/360x450_q75.jpg.

answers in an ever-shrinking window of time. The requirements of time-series data and the need for near-real-time response to detectable patterns are becoming greater, especially with the growing number of sensors being deployed within the smart grid. Unfortunately, developing a coherent business strategy for analyzing big data is lagging behind widespread experimentation projects as stakeholders try to "get a handle" on what Hadoop and other solutions can do, resulting in poor results that get pushed into production. This chapter addresses the important and salient issues of big data processing, what kinds of tools are available to help, and how to choose the right rigging to address business needs.

11.2 We Need Some Answers Here

When it comes to extracting value from big data, the landscape of choices is confusing. Most utility stakeholders come to the table hoping that big data will naturally provide better, faster, more-accurate answers to their questions. It does seem straightforward—when big databases are implemented that contain all of the utility's most important information, someone just needs to send a command to the database, and the system will just provide a useful answer. IBM's Watson may be able answer Jeopardy! questions, but it can't navigate most utility systems. At least not yet. The truth is, with big data analytics, the key is *not* to query with the presumption of a correct answer, but instead to *explore* the system in the hopes that something interesting will emerge. When that interesting thing surfaces, further investigation can be done.

At first, it's difficult to understand how this approach is at all useful when utilities really need measurable return on investment (ROI) with rapid understanding of the operation, especially when the utility is rife with shifting dynamics and demands high-value decision-making. How does this vague searching help anything?

Here's a story that might help: In 1854, James Perrott placed a bottle for hikers to insert notes into when they found themselves on a remote northern moor in Devon, England. The idea caught on, and trekkers began dropping postcards (sometimes addressed to themselves) in letterboxes and picking up fellow travelers' postcards, which they would post to their final destinations. Thus, *letterboxing* came into being. Eventually, the pastime became an art and letterboxes were so well hidden on this remote moor and throughout the world that they required clues to be located. These days, instead of postcards, searchers gather stamps from the treasure boxes. Some letterboxes even contain clues to the next letterbox. Much like the game of letterboxing, big data exploration requires navigational skill to find the answers. For some, it's a professional hobby; for others, it's a puzzle; and for others still, it's an art. Here is a letterboxing clue:

> From the ring of fire, the box awaits
> Take a few steps, magnetic north
> Do not be shy, step up and walk forth!
> From within a crack, home to a tree,
> Look under a rock, and the box will come free.

— Excerpt from Temple of Terror letterbox clue[2]

Data scientists are doing the exact same thing: Big data letterboxing requires adaptive procedures (hiking around), assistive algorithms (clues in the box), and queries ("Where do I go next?") to lead to the desired result—one clue at a time. Stratos Idreos describes this process of big data exploration as "adaptive":

> The system and the whole query-processing procedure [is] adaptive in the sense that it adapts to the user requests; it proceeds with actions that speed up the search toward eventually getting the full answer the user is looking for.[3]

Various vendors implement this methodology in different ways, but a true data analytics package is adaptive in some manner. This exploratory approach is only possible with newly evolving, highly capable, and fast query-processing systems. Even so, regardless of the technology used to provide these fast query times against some type of big data storage paradigm, the core challenge is the same: When big data is constantly being sucked in from a variety of sources, and when fast response is necessary based on that data, pat answers are not possible. And, in fact, unlike with traditional approaches to data where factual answers are expected from factual questions, with big data, it is impossible to comprehend all the relevant information that might be stored in the system (because it keeps coming and piling up in the system). Thus, big data processing must make a major departure in processing that is driven more from how the data will be explored, and not with the intention of fulfilling a request based on a preponderance of all known information.

11.2.1 How Long Does This Take?

With an optimized system, big data insights can occur faster than the time it takes to get a cup of coffee. By shifting from the notion of obtaining

[2] Green Tortuga (2009), "Temple of Terror Letterbox." *atlasquest.com*. Retrieved January 24, 2014, from http://www.atlasquest.com/boxes/clue/index.html?gBoxId=6.
[3] Stratos Idreos (2013), "Big Data Exploration." In *Big Data Computing*, edited by Rajendra Akerkan. Chapman and Hall/CRC Press, 273–294.

"information" to gaining "real-time insights," where real time equates to optimal efficiency, the utility will achieve economic value. Anyone who has ever worked with a lot of data has implemented the "submit and pray" model while waiting for the system to parse and decipher the necessary data. We work with what we have. Until now, really reliable and fast systems have been reserved by high-value use cases, but the data explosion and the declining costs related to advanced systems have changed that. Today, "fast data" is widely available, and the most common use cases include social network monitoring, sensor data networks, and high-frequency financial trading systems.

"Fast data" is another term that is best defined relatively. It's a way of delivering the right data at the right time from big data. Tony Baer of Ovum, who first used the term, describes it in practice as, "[comprising] a spectrum of technologies leveraging high-performance, multi-core processing, often in conjunction with silicon-based storage."[4] If you're a traditional business intelligence user who has tried to answer a critical business question in a timely manner, the need for this spectrum of fast technologies makes complete sense. Consider the scenario: A dangerous ice storm is coming, it's colder than it has been in decades, and a problematic situation is emerging in terms of excessive demand and the prospect of both planned and unplanned outages. The CEO has questions about handling the emergent issues. She needs to drive public communication and ensure that customer outages are contained and that a plan is in place for rapid restoration. Unfortunately, by the time the answer arrives from the analysis team, the ice storm has arrived and the crisis is in full swing. The process of obtaining answers that are detailed enough to drive specific action is too cumbersome. Once again, the utility is left with the prospect of trying to make good decisions for its customers in broad strokes with imprecise data, relying on experience and luck.

With poor situational intelligence on the grid, outages and disruptions last longer than they need to, and the impacts are sometimes unexpected. In January 2014, when the southeastern United States was hit with an extreme cold shock, an interruptible contract was invoked to decrease demand from a Kentucky university. This contract could be invoked with as little as a five-minute window of time to counter to grid stress. The storm was originally not expected to be so severe, nor drive such high levels of demand. Instead, it was one of the worst in the public power company's history. At the university, when the power was cut to the campus, the backup genset malfunctioned, and as students were being moved to warmer shelter, 40 percent of the campus's major buildings were

---

[4] Tony Baer (2012), "What Is Fast Data?" *Ovum*. Retrieved January 25, 2014, from http://ovum.com/research/what-is-fast-data.

suffering significant damage from frozen pipes and equipment.[5] The effort to balance supply and demand had severe unintended consequences.

How could this situation have been improved with the use of big data analytics? There are several possible approaches, but an obvious application comes to mind: A fast data-based analytical system that integrates weather data into the larger business decision model would have been a great advantage. By processing the best forecast models with historical utility data to generate a suite of probable scenarios, the utility would have had much more-precise information on the possible impacts to the grid. With those scenarios in hand, the utility could have better allocated and deployed assets to minimize the impact on the electrical grid's customers in a proactive and efficient manner, rather than reactively and in an incredibly short period of time for the customer (five minutes). This would have increased preparedness time (to enable spinning up the generators to see if they worked) and, in the case of malfunctioning equipment, helped to assure human and property safety in adverse conditions. With better, more-precise communication from the power company, the university could have avoided significant duress by bringing up the generators sooner, possibly avoiding or responding to the malfunction or making alternative arrangements with the utility to protect the students on the campus and preventing widespread damage to the facilities.

Even if the utility has access to lots and lots of the right data, the problem with utilizing large data sets is really not a storage capacity problem; it's slow indexing, tuning, and data access speeds. With traditional data-processing technologies, there are inefficient choke points between storage, processing, and querying. As discussed, the manner in which big data must be approached is very different than traditional data problems, which makes it is difficult to compare the "old" process to the "new" process (and the subsequent outcomes). In fact, it is necessary to completely break free from customary perspectives of how data is managed and accessed. This is the first step in working to implement future-proof big data analytics architectures, and it is an approach largely held back for psychological reasons ("But this is how we've always done it.").

This is not to say that there isn't tremendous value in the traditional enterprise-oriented data warehouse (that single source of truth). In fact, it's likely that these implementations will continue to be valuable for quite some time. They just have to be utilized for the right kinds of business problems.

---

[5] Rob Canning, Chad Lampe, and John Null (2014), "Damage to At Least 40% of MSU's Main Buildings Due to Power Outage and Freezing Temps," *WKMS*. Retrieved January 25, 2014, from http://wkms.org/post/damage-least-40-msus-main-buildings-due-power-outage-and-freezing-temps.

However, it must be acknowledged that big data analytics are challenged by different forces—in sourcing, storage, and usage. It is often an endless and unpredictable stream. The customary approaches do not comprehend an influx of continuous data inputs that are needed for many forms of real-time analysis. For example, in the utility, these "infinite" sources of data include grid sensor data, monitoring information (including video), energy-related commodities-trading functions, inputs required for unplanned outage recovery, and some asset health applications.

11.3 Mining Data for Information and Knowledge

As early as the 1960s, statisticians began disparagingly using the terms "data dredging" and "data fishing" to refer to those who fiddled around with data without any sort of a priori hypothesis about their results. By 1990, the term was repurposed by many in the data community as "data mining" to refer to the sort of archeology we now depend on in extracting value from big data sets. However, this mining process itself has been in use for centuries. For example, regression analysis, which is a statistical process for estimating the value of one variable from the values of others, was employed as early as 1794 by German mathematician Carl Friedrich Gauss in the application of the "method of least squares" (who actually invented the method is debatable, though its roots likely are found in the poetic ideals of the ancient Greeks).

Putting the actual strategy of least squares aside, the method evolved from a problem that is quite salient to big data analytics: when sailors can no longer rely on the horizon for navigation, they dredge the skies. Thus, an accurate model is required to define the position of celestial bodies for navigation (which is just not easy, given the inconvenient curvature of the Earth, the shapes and sizes of celestial bodies, and their trajectories). Thus, using the model, Gauss showed that there is an arithmetic approach to consistently locating these bodies by minimizing the errors found in estimation. And, similarly, data mining is a way to programmatically extract patterns from data sets that creates new information for further use.

A very common example in the utility industry is found within the many programs that attempt to advise property owners on how their homes or buildings compare to similar structures in their neighborhood or region, and how they can improve their own structure's performance. By analyzing consumption data and information obtained on the structure, such as building-envelope data, the utility does not have to go into each building to identify leaks or inefficient heating and cooling equipment. It is able to use a statistically derived benchmark to measure expected savings from improvements performed on the structure, and thus provide a service to the customer where it recommends and prioritizes applicable improvements that can have a measurable impact to save energy, money, and help the utility drive down overall demand.

This type of machine-learning is also a very powerful tool for predictive analytics that allow applications to predict asset failures and outages, detect revenue leakage and theft, and identify optimization opportunities on the grid. The increasing penetration of distributed renewables on the grid drives additional valuable use cases, where data mining can ease the integration of intermittent generation by creating opportunities for data-driven decision-making, managing unpredictable generation, and lessening the negative impacts related to voltage disturbances created when microgrids island and reconnect.

Data mining is the key to understanding a vast collection of facts by discovering the right associations and relationships among those facts to move the utility toward the knowledge that not only describes the past and predicts the future, but allows the organization to take appropriate action. That sounds a lot like data analytics. However, there is a distinction between data mining and data analytics; largely, that data mining is focused on discovering hidden relationships, and data analytics is focused on deriving some conclusion based on known information. Many argue that this is a false distinction. It is well within the scope of descriptive analytics to discover patterns. It is also well within the scope of predictive analytics to discover a pattern that describes customers who are at risk of defaulting on their bill, and finding customers who may also fail to pay in the future. It is fair to say that data analytics is just a new name for data mining; although, it may be more accurate to conclude that data analytics is an extension of the practice of data mining. For example, while a data-mining exercise might expose that most of the customers who eat lunch at Whole Foods just came from yoga class, it is data analytics that will help Whole Foods decide what to do with that information, such as offering smoothie coupons pinned to a point-of-sale yoga mat, or even expanding its marketing efforts to target those who cannot touch their toes. Or, to be true to our first example of regression, data mining might describe the celestial bodies, but it is analytics that determines which customers will be most interested in either the admiral's sextant or a lifeboat.

11.4 The Process of Data Extraction

If we agree that big data analytics have expanded the existing approaches to data mining to more extensively work with the data beyond simple pattern recognition with statistics, it is important to understand what techniques are most useful, prevalent, and evolving. Firstly, it is critical to achieving measurable ROI to

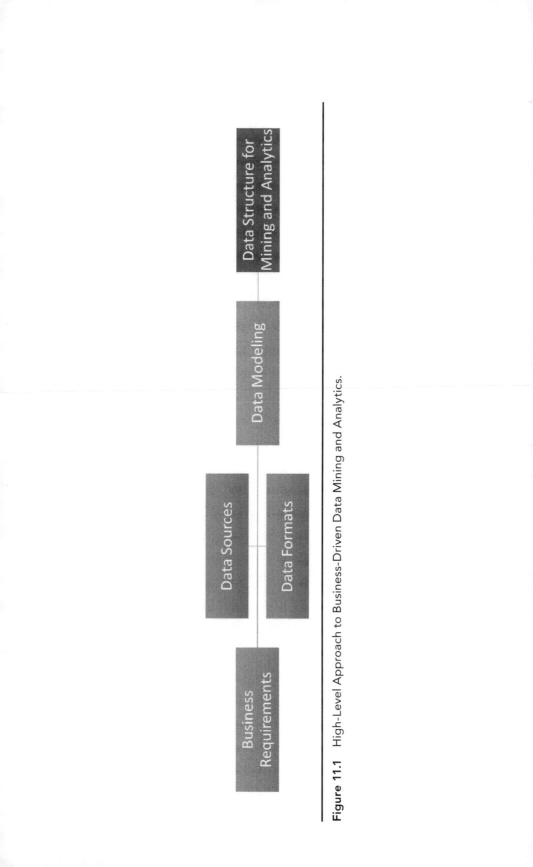

Figure 11.1 High-Level Approach to Business-Driven Data Mining and Analytics.

drive the analytics program directly from an honest and thorough assessment of business problems and objectives.

Like most complex processes, this approach is often iterative as the utility begins to identify the different data that *needs* to be extracted and that *can* be extracted. This starts with identifying facts that will help treat targeted issues, and then looking at the source data to best associate and cluster with other data to produce the desired outcomes. Figure 11.1 describes a high-level approach to creating a business-driven data structure that can be mined and analyzed for deeper insights. Depending on a variety of factors, and regardless of the techniques ultimately used for data analysis, how the data is sourced and its format affect how the data is stored, processed, and described through the data model. As you will recall, the data model must faithfully represent something that can be structured (sometimes called "real-world" objects). Thus, the correct rules and concepts used to define those models are those that will most precisely define the object in question (for example, a network of sensors or a customer).

In building models for processing and extraction, there are some key techniques that are used by a number of different tools. Unfortunately, many data solutions vendors do not share terminology and sometimes a buzzword will nudge its way in, increasing confusion and perceived complexity. There are some very basic techniques that form the foundation of more-advanced analysis. Although the enumerated techniques are not in any way exhaustive, these terms are provided as a foundational primer on some of the workhorse algorithms used in analytics processing.

Association (or relation). This is what most people will immediately think of when they consider pattern identification modalities. This technique simply allows the data scientist to explore correlations between two things to identify patterns. These are the kinds of information that support a variety of business functions, such as marketing, inventory management, and customer relationship management. An example application that uses the associative model relates to selling retail products to customers whose appliances might show certain load patterns, such as an air conditioner that is exhibiting detectable signs of imminent compressor failure, like hard starting or motor overloading. The utility can associate the behavior of the unit with the need for a repair or a new unit, offering services or discounts to customers that will encourage them to repair or replace their appliance before it nosedives, simultaneously driving the opportunity for a more-efficient replacement. Essentially, the use of association helps find rules that can be used for cross-sell, root-cause analysis, and defect analysis.

Classification. Also a very familiar technique, classification is the method used to generate an idea of a particular customer, item, or other object by describing its attributes for the purposes of classification. Describing cars (how many wheels, number of seats, and so on) is the classic example, but perhaps a more useful example is of children playing with Lego blocks. They have a certain structure they want to build, but they need to know what kind of pieces they have. So they sort them based on various characteristics—usually brick size and color—to assess their available resources. This example shows how classification results can be driven into other techniques, especially clustering. In general, classification is often used to help predict outcomes very quickly and simply, using algorithms based on conditional probabilities and scoring.

Clustering. Used to identify natural groupings, clustering creates assemblages, where, not surprisingly, the members of a particular cluster are more like each other than the members of a different cluster. Clustering is also used in more-complex analyses based on *nearest neighbor* data. The nearest neighbor is a form of establishing identity based on the notion that if a certain structure shares attributes with other structures in the same cluster, they're likely to share other attributes. Any classifiable case benefits from clustering analysis not only to recognize patterns of data without requiring an exact match to any other known pattern, but to identify new, previously undetected cluster relationships.

Decision trees. A decision tree is a graph that models the preponderance of tests and consequences, which may include chance, costs, and usefulness. These graphs are easy to understand but can quickly become quite complex when values are uncertain or when many of the outcomes are linked together. Decision-tree analysis is a fundamental underpinning of asset management where the basic questions are of repair, refurbishment, replacement, or even augmentation. Without analytics, asset decisions are based on an end-of-life prescription or a rule-of-thumb analysis, meaning some assets are replaced too early in their effective life, or are run to failure. This is hardly optimal. A decision tree can be used that takes into account risk-weighted economic costs, condition, performance, business risk, and the interaction of all these factors. Although the most rudimentary models may be substantively qualitative, asset management applications in utilities benefit from sophisticated modeling approaches that tie intervention to objectives, as opposed to the avoidance of events or negative scenarios.

Feature selection. By combining existing attributes, feature selection produces new attributes. One important approach to feature selection is principal components analysis (PCA), which helps find patterns in data of high dimension. PCA is a statistical analysis that exposes the internal structure of a certain data set that helps explain variance in the data.

Sequential pattern. Sequential pattern analysis refers to the use of algorithms applied to longer-term data sets to identify trends or repeated events of a similar sort. The ability to cull out sequential patterns can be quite valuable for strategic decision-making applications because of the ability to detect events, identify anomalies, and make predictions. This kind of analysis is closely related to time-series data analysis, as both approaches examine discrete values that are delivered in a sequence. Failure prediction is another interesting application of sequence mining that can play a crucial role in detecting system failures by identifying the frequency of events or set of related events that can be utilized by a prediction system to be context- and time-aware.

11.4.1 When More Isn't Always Better

Big data proponents will almost always argue that more data trumps better algorithms by allowing the data to speak for itself. And it's true that more data will almost certainly provide the opportunity for greater prediction accuracy, but the viewpoint is overly simplistic. The bottleneck in big data technologies is not communications latency, slow processors, or thrashing hard disks. Hardware and storage paradigms have improved sufficiently that the real problem that data analysts are facing is finding the right software that will make sense of the data—specifically, which of all the data should be analyzed and how should it be analyzed to make any sense of it. The strength of the algorithm must complement the manner in which the data is extracted and processed in order to realize the benefits of running more data.

A classic example is the Google PageRank algorithm. Early search engines processed the text of web pages to produce search results. In 1998, Google tweaked the traditional algorithm to consider additional data in the form of hyperlinks, and it weighted the text within the hyperlinks almost as heavily as the page title. This is not algorithmic genius, but it's a very important concept: Adding more data from varied sources is usually better than designing a new algorithm. However, there is also a tipping point, where too much data is overwhelmed by noise. Simply think about how your brain works in a conversation. Most of us are taking in many inputs, including words spoken to us, smells, body language, and facial expression. If we try to memorize every word because we believe that's the only important data, we are algorithmically filtering out other important information that will help us negotiate the facts and even the relationship itself. Just processing the words from the conversant will not make it easier to understand the overall context of the interaction (that would be a bona fide data deluge). However, being able to connect the words with emotions and other aspects of the exchange builds meaningful understanding. Similarly, with big data value extraction, the algorithm is designed to enhance the understanding of the data by drawing connections between many data points; it isn't just the raw preponderance of bits.

11.4.2 Running for Performance

While there are plenty of vendors who will argue with this point, it seems quite clear by now that many facets of big data cripple legacy processes and tools. Big data analytics increase computational significance, Structured Query Language (SQL)-based algorithms are often inflexible, and it is antithetical to analytics to lock into a specific platform, inhibiting quick evolution and the integration of new data and tools. In searching for the right solution, based on the business problems and data at hand, stakeholders must look for raw speed, the ability to scale out (not up), functionality, compatibility with other applications and systems, easily managed platforms, and—most importantly—simplicity. There is an important caveat, though: No technology solution—especially open-source software like Hadoop—stands still. Change is inevitable.

11.4.3 Hadoop: A Single-Purpose Batch-Data Platform?

Well, maybe. In the previous chapter, we introduced Hadoop and MapReduce for storing and aggregating big data. But not surprisingly, given the wild pace of innovation and technology evangelists, the young Hadoop may be insufficient and constrained from treating many forms of data. This fact may change quickly, but the rate of innovation may be causing a state of paralysis in the utility industry that's trying to make complicated choices about how to proceed in bringing up a data analytics project.

Yes, there are limitations for processing with MapReduce—currently, it is batch, and many customers require stream processing. Until the general availability and stabilization of Hadoop 2.0, with its new architecture that marks an expansion of the singular focus on MapReduce toward other processing patterns, MapReduce is a poor first choice for low-latency processing of fast-stream data algorithms. These algorithms are especially beneficial to power-sensor data analytics, which are the heart of the smart grid. In fact, in 2010, Bill McColl PhD (founder of Oxford Parallel) said, "[B]atch processing tools like MapReduce and Hadoop are just not powerful enough in any one of the dimensions of the big data space. ... Hadoop is great for simple batch processing tasks that are 'embarrassingly parallel,' but most of the difficult big data tasks confronting companies today are much more complex."[6] At the time, McColl obviously didn't anticipate stream-process capabilities in Hadoop. Sensor and machine data processing are now a broad reality, and it appears that the project is adapting.

But until it is proven that Hadoop can handle stream data, the primary Hadoop programming framework of MapReduce is its primary style, a style that represents a form of dealing with data that simply cannot solve every conceivable problem. Even if it is economically advantageous (read: tempting) to store smaller data sets in Hadoop Distributed File System (HDFS), it has been shown to actually be slower in dealing with smaller data sets than other applications. Hadoop is best at processing massive volumes of data using MapReduce. At the moment, the use of MapReduce only makes sense when the files being operated on are large and rarely updated or appended. For example, it is not a good choice to use MapReduce within the customer service operation where lots of changes are being made to fairly constrained sets of data. On the other hand, it's very well suited to serving up analytic queries on a command designed to discover every user whose consumption patterns are most similar. Similarly, Hadoop has limited value in online environments that depend on rapidly processing small amounts of data.

The appropriate use cases for Hadoop are important to understand within the context of the utility. In the world of big-grid data, much of the information is delivered asynchronously and in a variety of formats, and a batch process that prefers static files, such as Hadoop (no matter how fast you can make it), may not be the right paradigm for streaming data for quite a while. Hadoop is powerful, but perhaps the hype is even more powerful. It is up to the appropriate utility stakeholders to determine where Hadoop best fits within the organization to understand which analytical functions it can support, which it cannot, and how it is continuing to evolve as a technology.

11.5 Stream Processing

Stream processing is a programming paradigm that is key to the overall big data value extraction process, as it supports the ability to analyze data in motion—that means accessing the data to derive value and find relationships among the

---

[6] Bill McColl (2010), "Beyond Hadoop: Next-Generation Big Data Architectures," *New York Times*. Retrieved from http://www.nytimes.com/external/gigaom/2010/10/23/23gigaom-beyond-hadoop-next-generation-big-data-architectu-81730.html.

data points before the information even hits a disk. This form of data processing is well matched to computationally intense applications and goes far in meeting the demands of data that is being continuously fed from data source to data consumer. If an organization has a business requirement to quickly analyze measurement and event data in both structured and unstructured forms as it arrives, stream processing is the best way to obtain these real-time insights.

Unlike with conventional database management systems (DBMSs) where the database analyst might execute some query against the database, with data stream management systems (DSMSs), the same query is continuously executed against a volatile data stream. Specifically, a DBMS assumes the data is in an exact and accurate form, while a DSMS is designed assuming that the available data may be outdated and even inaccurate, and adapts to those deficiencies. The DSMS is data driven, which means as long as new data arrives within the system, the query will produce new results.

11.5.1 Complex Event Processing

One example of stream processing that's well known to the utility is complex event processing (CEP). CEP is a form of event processing that combines multiple data sources to detect events or patterns of circumstances.

Imagine standing on a street on a chilly Thursday night in downtown Boulder, Colorado. You hear the familiar strains of "Happy Birthday" coming through the steamy restaurant glass across the way. At the end of the out-of-key singing, you hear clapping, cheering, and the squawking of noisemakers. Aha! You cleverly determine that someone just became a year older! To determine this, you analyzed various inputs and correlated the events to perform a personal act of event pattern detection; similar to one of the important things that CEP (sometimes simply called event stream processing) does for the utility.

CEP analyzes streams of data from a variety of distributed systems on the smart grid with the goal of combining the data to infer patterns that can suggest the nature of an occurring event. The best use of CEP is to analyze historical time-series or streaming data that is on a time continuum Many systems are also designed to automatically trigger a response to the conclusions drawn by the CEP when the need to act upon live market conditions is required.

There are three fundamental use cases for CEP in the context of the electric power business:

1. Identifying critical situations through known event patterns
2. Detecting signals that may lead to new opportunities
3. Detecting and identifying important changing conditions

The monitoring and detection of system state on supervisory control and data acquisition (SCADA) networks is a key application of CEP for grid management. But there are other, more surprising utility-related applications, including demand response (DR), that benefit from CEP. One example use case is the sensing of data from commercial buildings, where CEP applications can detect disconnected devices and meters, transgressions of program thresholds during events, and nonconforming curtailments after a DR event and during the measurement and validation (M&V) step of the DR workflow.

Event-driven systems are not new to the utility. However, the advancement of analytics and affordable processing is driving the ability to meet a plethora of undertreated business problems and opportunities. The ability to quickly apply rules to atomic events on a stream afforded by CEP technology is resulting in a continuing growth trajectory for the technology that is well positioned to meet any sort of surveillance system, from grid pattern processing to financial-trading applications.

11.5.2 Process Historians

When it comes to data, one walk around almost any utility and you're sure to trip over a process historian. Sometimes referred to as operational or data historians, process historians are end-to-end solutions that manage real-time data collection, archival, and distribution of time-based process data within a centralized system for both real-time and historical views to users across the utility enterprise. Thus, on a single platform, the ability to historize, search, analyze, and access is self-contained. The process historian is purpose-designed to capture and manage plant information that includes status, performance, tracking, compression, security, and presentation. Early in their evolution, these devices were largely focused on plant operations, but with a growing interest in operations and implementing efficiencies in the entire process, these systems are now the incumbent solutions delivering their data to information technology applications within the enterprise from a variety of data sources and control networks.

Clearly, the process historian has many benefits if it can be depended on for optimal performance and the appropriate and desired tools for data analytics are available; however, there are drawbacks from the analytical perspective. The predominant systems in the utility are focused on key performance indicators (KPIs), and thus are more concerned with metrics and compliance than true data science applications. But solution providers are adapting and partnering to help drive the kinds of analytics that the industry demands, supporting operational management and advancement. The continuing success of these solutions will be their ability to expand their partner ecosystem to help encourage the collaborative use of operational data for the entire enterprise, above and beyond a limited focus on operational systems.

11.6 Avoid Irrational Exuberance

Big data interest and exuberance are growing as utilities work to determine how to best harness the relatively untapped resource of the surging volumes of data. Unfortunately, implementation and deployments of Hadoop and other data frameworks are gaining investment, while big data strategy is not. Add to that the growing real-time delivery of analytics and the prediction that by 2015 fruitful data analysis will depend on the successful ability to operate on real-time data.[7] While IT staffs are implementing Hadoop on relatively small data sets for testing and validation, they may be using exactly the wrong tool for the problem.

Utilities are under pressure to adopt big data tools; however, the landscape can be so confusing that some will be very slow to adopt or spend a significant amount of time in an evaluation phase. Most utility managers are not IT savvy and therefore will not be completely comfortable understanding the big data analytics needs of the organization. Support is required, including in choosing applications and tools that meet the scope of user demands.

Next Steps
• Determine the high-value areas of interest within the utility
• Seek out peers and case studies that illustrate how adjacent industries and other utilities are tackling these big data problems
• Ask vendors to present their strengths and challenges; summarize those findings
• Ensure that your approach considers multiple options and vendors that are best suited to meet the characteristics of the sourced data and the demands of the users
• Plan for future needs

---

[7] Intel IT Center. (2012), Peer Research: Big Data Analytics. *Intel IT Center | Peer Research*. Retrieved February 01, 2014, from http://www.intel.com/content/dam/www/public/us/en/documents/reports/data-insights-peer-research-report.pdf

Chapter Twelve

Envisioning the Utility

Astronaut Edgar D. Mitchell, Apollo 14 lunar module pilot, reads a map as he moves across the lunar surface during extravehicular activity. (*Source:* NASA[1])

12.1 Chapter Goal

The business value of an analytics program is severely devalued if the utility is inhibited from making better decisions in an increasingly dynamic environment

---

[1] Image retrieved from the public domain at http://www.dvidshub.net/image/837746/mitchell-studies-map#.UvlxBr_COxN.

because the information presented doesn't make sense. This chapter introduces the basic concepts of data visualization and describes how, due to the way humans process information, data visualization may be one of the important answers to the question "What are we going to do with all that data?" We discuss the foundation necessary to develop a sense of visual literacy and an understanding of when to incorporate visualization strategies and how they can be beneficial in all aspects of the utility business.

12.2 Big Data Comprehension

It's safe to say that the future of energy delivery is about innovation and discovery. As a process of exploration and data interaction that drives high-value action, big data analytics are the foundation of this innovation. Visualization-based tools are advancing quickly to support this process and allow both operational and business users to pull together disparate data sources (in what is often called a "mashup") to create custom views that support highly customizable and relevant analytics. Additionally, mobile devices such as tablets, netbooks, and even smartphones now carry enough onboard graphics power to make intuitive, visualization-based data discovery tools available to multiple users across the enterprise. While data scientists have an important role in designing powerful and accurate models for the utility, the ability to explore data and draw actionable conclusions is becoming democratized. Indeed, if properly governed and secured, this democratization has the potential to drive down operating costs and drive up innovation.

However, visualizations are not inherently helpful—in fact, they can be confusing and misleading. The utility needs to incorporate visualization technologies that do more than just describe current state; they must help the utility predict emerging conditions on the grid, reveal hidden relationships that introduce new efficiencies, and provide stronger decision-making capabilities.

Data visualization strategies are quite varied, and though they may be tuned to support the underlying data classes, the best visualization tools are the ones that will help their users readily home in on the subject of their analysis. There are several general characteristics that describe a comprehensive data visualization tool, including:

- The ability to work with a real-time data stream
- Support for multiuser collaboration
- Fast processing time
- The ability to export analysis for reporting

Other features that may be important are the ability to access some subset of the information on a mobile device, touch optimization (especially for workforce applications or operators), and—importantly for the utility—governance features that provide a chain of custody for data lineage and user operations on the data.

Big data analytics in the utility will serve the entire enterprise and touchpoints across the meter demarcation—from business and operations to customer service, field operations, and the energy consumer.

12.3 Why Humans Need Visualization

But how exactly is the big data industry delivering big data and realizing the unprecedented promises of discovery, collaboration, and exploration? Those are grand promises, to be sure. So how do we get there, get those insights, and hope to comprehend all the information we need in order to make better decisions?

Often in this book, we have attempted to provide real definition to important terms. And visualization is all about conjuring the universe in a new way. That surely doesn't sound precise, but once again, we need only consider the wisdom of the ancient Greeks to realize the simplicity of this statement. It is something, as human beings, we do in every moment of our waking life. We use shapes, sizes, and positions to classify information. This conjuring was documented by Aristotle on the topic of logic found in the text *Categories* from his collection *Organon*, where he says:

> Of things said without combination, each signifies either: (i) a substance . . . ; (ii) a quantity; (iii) a quality; (iv) a relative; (v) where; (vi) when; (vii) being in a position; (viii) having; (ix) acting upon; or (x) a being affected. (*Cat.* 1b25–27)[2]

Categories is indeed the foundation of many philosophical approaches, especially in the area of science where we pursue the ability to understand our complex universe with some sort of comprehensible categorization. Consider the ripened tree, with its roots, trunk, branches, leaves, and fruits. A tree has an internal order to it and is the basic structure of many of our most useful ways of representing information, including hierarchical depictions and graphical storytelling. And it remains one of the most explicable approaches

---

[2] Christopher Shields (2008), "Aristotle," *Stanford Encyclopedia of Philosophy*. Retrieved from http://plato.stanford.edu/entries/aristotle.

for understanding complex systems, as we build from the roots to the branches and extend to the fruits.

But unfortunately, our worlds are not as compulsively well ordered as a tree structure would lead us to believe, and anyone who has ever examined the swarms of wires, sensors, and devices at a utility knows this well. Instead, as we pile on more intelligence and integrate new forms of generation, we find a structure that is not nearly so neatly centralized, organized, and classifiable. Despite our best intentions (and hopes), the grid doesn't work that way anymore, and this reality is a quagmire for the utility stakeholder attempting to do something as seemingly straightforward as balance the load. It's indeed a fact that as the hierarchical order of the grid diminishes, the days of tree-structure logic are slipping away.

12.3.1 Walking Toward the Edge

Moving from our well-understood tree structure to accepting the need and value of visualization brings us to what's known as the "problem of seven bridges." As the story goes, there was a puzzle the townsfolk of Königsberg, Prussia (now part of Russia), entertained, which was the question of whether it was possible to walk through town and visit each part of the village, but cross each bridge only once. As you can see in the map shown in Figure 12.1, Königsberg spanned both sides of the Pregel River (the town was decimated by bombs in World War II) and included two large islands, and seven bridges crisscrossed the city. At the time, a Swiss mathematician named Leonhard Euler (1707–1783) was working at the Berlin Academy in Germany, where he was presented with this very problem in 1736. The rules were that each bridge would be crossed only once completely (no retracing and no halfway crossings), but it was not necessary to start and end the walk at the same spot.

Euler realized that attempting to list all the possible pathways would be way too exhausting and maybe impossible, so he abstracted the problem to consider only the landmasses and the bridges. Today, we would call the landmasses "nodes" (or vertices) and the bridges "edges," and the result is the basic vocabulary of graph theory. The problem could be solved on this new topological structure by taking a Eulerian walk—a map of nodes and edges where the connection information is the only relevant aspect to the problem. Thus, with the help of the graph, for every node entered on an edge, it would have to be left by another edge. So, to solve the Königsberg issue, the number of times one enters a nonterminal landmass must equal the number of times one leaves it to cross a different bridge. If every bridge has been crossed exactly once, then each one of those landmasses must have an even number of bridges (for coming and going). They didn't. So the disappointing answer to the bridge problem? No.

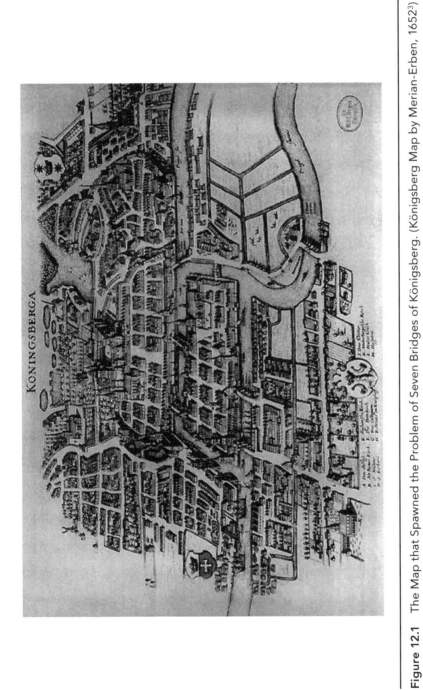

Figure 12.1 The Map that Spawned the Problem of Seven Bridges of Königsberg. (Königsberg Map by Merian-Erben, 1652[3])

[3] Retrieved February 12, 2014, from http://commons.wikimedia.org/wiki/File:Image-Koenigsberg,_Map_by_Merian-Erben_1652.jpg.

Why are Euler and the origins of graph theory so important to our current topic? They help us understand that no matter how complex the combinatorial problem at hand may be, it can be abstracted so that extremely hard data problems can be solved based on how nodes are connected to one another—the very foundation of network science. By observing nodal relationships spatially, in a manner that is unaffected by the shapes or sizes of the nodes themselves, we filter out irrelevant information that gives human cognition a boost. Based on a question of modest beginnings, Euler's approach to solving the problem now allows us to develop very powerful models that enable the facility to predict and optimize all varieties of networked systems, including the Internet, telecommunications networks, the electricity grid, and psychosocial systems.

12.4 The Role of Human Perception

The notion of distilling the important pieces of information required to solve the problem, as Euler identified in answering the question of the Seven Bridges of Königsberg with the invention of the graph, is a very important concept in constructing useful interfaces that support human perception. While not an application of discrete mathematics directly—like the graph—a successful visual tool will use approaches that meet the needs of the human visual system by focusing attention on the area of critical need. This can be accomplished precisely in the approach that Euler took: by ignoring the shapes of the landmasses or the lengths of the bridges in developing his methodology.

The reason that visual constructs are so useful is that they maximize the efficiencies of the brain. Very simply, we see things quickly, but when we have to think about them (cognition), it takes more time. Thus, the visualization of big data provides an opportunity to rapidly and comprehensively understand the underlying data. And sometimes, it can even simplify the data without undermining the key value within the message. How to produce useful visual artifacts and tools is a discipline with devotees of various stripe, although the approach is almost always the same. To get from data to understanding, we must be clear on the characteristics of the underlying data, as well as the right way to visualize that data for maximum effectiveness and identify the rules that map the data to the visual representation.[4]

---

[4] Jessie Kennedy (2012), *Principles of Information Visualization Tutorial—Part 1: Design Principles*, Institute for Informatics & Digital Innovation, Edinburgh Napier University. Retrieved February 15, 2014, from http://mkweb.bcgsc.ca/vizbi/2012/principles.pdf.

Graphic visualization is really nothing more than a system to convey information, a system that attempts to normalize the inputs from industrial designers, computer scientists, political scientists, cognitive psychologists, ethnographers, statisticians, and artists. Luckily, there are design principles that form the foundation of how to drive an accurate perception by bringing together the pieces of data to create a useful whole that avoids ambiguity. Many of these principles are drawn from what we now know about how the brain processes visual information.

12.4.1 Preattentive Processing

Preattentive processing has been investigated for many years by researchers seeking to understand how human beings analyze images. With the rise of big data visualization, this field has once again emerged as an area of topical interest. There are certain visual properties that can be detected very quickly and accurately by the visual processing system. These preattentive tasks are completed so fast—in less than 250 milliseconds—that they might be considered "intuitive." Just for context, an eye movement alone can take 200 milliseconds to initiate, yet even in such a compressed time frame, it's possible to focus a person's attention with ease.[5]

Consider Figure 12.2 for a moment. Without any special effort, it's very likely that it was an instantaneous act to identify the presence of a bigger circle

Figure 12.2 Illustration of the Property of Size on Preattentive Processing Capability.

[5] Christopher G. Healey (2009), *Perception in Visualization*, Department of Computer Science, North Carolina State University. Retrieved February 13, 2014, from http://www.csc.ncsu.edu/faculty/healey/PP.

in the group of other circles (called distractor elements). The unique property of the circumference of the larger shape allowed it to "pop out" of the other elements in the picture. If there were two circles that shared that larger circumference size, they would immediately become nonunique and would not be detected preattentively as a shared target, although you might have noticed right away that there were two unique objects. Clearly, if the graphical properties used during data visualization can draw the user's attention appropriately to the critical areas of interest in the display, there will be less interference, less chance for confusion, and increased speed and efficiency in understanding the impact of what is being presented.

Glanceability

It is critical to understand the forces of preattention when assessing and selecting visual analytic tools for a big data analytics program. This is perhaps especially important in the operational context, where discrimination between graphical features can confound and confuse the user—or worse, become completely meaningless—leaving operators more deficient than when they relied on streams of textual data and simple alarms. Some user interface designers call preattentive information "glanceable," or that which can be seen and understood in a single glimpse.

Preattentive features or graphic devices include:

- Color
- Orientation
- Lighting direction
- Size
- Closure
- Curvature
- Length

Features can be applied in a variety of ways and contexts that tap into these different processing tasks, which encode graphical elements by employing the following techniques:

- Position
- Length
- Angle
- Connection
- Slope
- Area

- Shape
- Containment
- Density
- Saturation
- Hue
- Velocity of motion
- Direction of motion
- Texture[6]

There are certain processing tasks that the brain performs during that glimpse, wholly dependent on the arrangement of the visual features, including:

- **Target detection.** A unique visual element among a field of distractor elements can be rapidly detected for either its presence or absence.
- **Boundary detection.** By creating collections or groups of common elements, where each group has a mutual visual property, natural boundaries are created between groups.
- **Region tracking.** When one or more visual elements are unique in nature, they can be rapidly tracked as they move through time and space.
- **Counting and estimation.** This task combines any number of elements that are unique in their visual features, which can be counted or estimated by the user.[7]

If the visualization is successful, the right information will be focused upon. If preattentive processing has not been controlled for, it is very likely that the important information will be missed, creating the potential for error and driving down user trust for the usefulness of the system.

As an example of the impact of various graphical devices, Figure 12.3 shows two different depictions of exactly the same data. On the left is a very simple example of grouping that creates boundary detection to show the 2010 national average (approximately 1 in 5) for the segment of people who are age 55 and over and are employed in the utility workforce.[8] Compare this image with the pie chart on the right, which uses a statistical graph instead of an infographic. At a glance, it is clear that the infographic image, while perhaps more visually appealing, does not show proportion well and does not demonstrate the impact

---

[6] Kennedy [4].

[7] Healey [5].

[8] Joshua Wright (2010), "Data Spotlight: More Than 1 in 5 Utility Workers Are Retirement-Aged," Economic Modeling Specialists International. Retrieved February 15, 2014, from http://www.economicmodeling.com/2010/06/29/data-spotlight-more-than-1-in-5-utility-workers-are-retirement-aged.

Figure 12.3 Hue and Boundary Detection Technique Compared with a Pie Chart for the Same Data.

of 20 percent of the workforce being at risk. The point is not to denigrate the infographic depiction, but simply to show how quickly we may size up a situation based on what our visual brain tells us—and how much power is in the hands of the visual designer.

The use of preattention features is just one of the ways in which visual designers tap into the functioning of the human brain for optimal information transmission.

12.5 The Utility Visualized

There are two reasons for using visualization as a tool for data analytics: to explore or to explain. Exploratory visualizations are of the utmost value in the realm of data analytics, and they help analysts discover new patterns, identify emergent trends, or find microproblems that call for further exploration. This type of tool is particularly useful for analysis on data sets where there may not be a deep understanding of the meaning of the content. In an operational context, exploratory analytics must be highly effective and promote efficiency; they are also extremely useful in exploring operational data streams to treat issues such as asset problems. Primarily used for communication and not really for analytics, explanatory visualizations may be combined with exploratory techniques to transmit key pieces of information or even a particular perspective on the data. And they're most valuable when a story about the data needs to be told, quickly and accurately.

Some big data vendors are tempted to include visualization with their products, although they sometimes seem to be doing their best to avoid well-founded and thoroughly researched design principles. It's not uncommon to see bizarre color combinations, distracting animations, and gratuitous graphics that actually increase the time and effort required to make sense of what's being presented. In fact, the problem has gotten so far reaching, that there is an entire website dedicated to bad visualizations (wtfviz.net) to cast humor on the issue, and many books have been written on the topic of how to lie with maps and infographics. However, in operational settings that can account for life or death, or when making costly decisions that will substantively impact the viability of the business, poor data translation can have severe consequences.

Surprisingly, the most egregious infraction within such data visualization applications is the display of information for what we already know. It's redundant and adds no value. An example is a dashboard where the metrics change slightly every day, but usually well after the point when anything can be done responsively. In fact, if anything dramatic had happened, such as a storm causing major outage issues, we'd know about it long before we sat down to view our

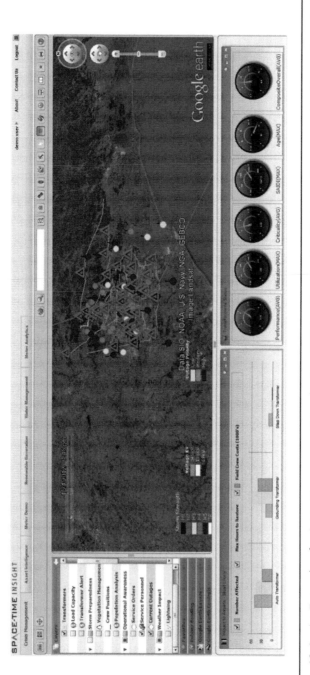

Figure 12.4 An Example of a Crisis Management Visualization Utilizing Preattentive Features. (Image provided by and used with permission of Space-Time Insight, 2014.)

daily executive dashboard. Instead, the true value from analytics rests in finding a view into new discovery—finding those things we didn't know about but really ought to be aware of if we have any chance of uncovering a new course of action. This is the purview of visualization.

When it comes to massive volumes of complex, variant data, the ability to analyze data that is in motion, or hasn't been preprocessed or rationalized, requires the users of this data to rapidly identify anomalies and outliers that are potential harbingers of trouble or other negative system and business impacts. A string of values where one might be colored red and another might be screaming #REF! from a data table is not going to achieve this, but visual metaphors that direct the user to important information and potential impacts will.

Many vendors are beginning to incorporate powerful visualization into their product offerings, and others are focusing solely on visualization tools. Space-Time Insight is best known for successfully deploying its geospatial and visual analytics applications in the California independent system operator (CAISO). The company focuses on delivering situational intelligence and provides an excellent example of conveying critical information using preattentive cues in an operational context.

As shown in Figure 12.4, multiple sources of data are correlated, analyzed, and presented to users in a single interface. In this case, the operator is viewing crisis management information and seeing transformers (represented by triangles) that are in the path of a storm, depicted by the line at the bottom of the map. The circles are the correlated outages. The chart at the bottom left shows the number assets affected, the estimated time to restore power, and even the cost associated with that restoration. The dials included on the screen show various factors correlated with performance and utilization for critical assets in play on the map. Each one of these risk measurements has been derived from a number of variables from disparate sources. Despite its simplistic look, the data represented on a single screen in a glanceable format is extensive.

Visualization techniques can be used in conjunction on a single display. Each has a role in helping the underlying data communicate a meaningful and understandable message, even when the number of underlying data points may be in the millions and several data sources may be mashed up to provide completely different classes of data. Edward Tufte, one of the most important writers on the topic of visual representation, identified many ways that graphical elements can go wrong, especially referring to a very common issue he called the "Lie Factor." Specifically, this factor is a value that describes a relationship between the size of a particular graphic and the size of the effect that actually exists in the data. That means that the graphic is way out of whack and either under- or overrepresents the truth of the matter. At the risk of severely misguiding an operator, optimal visualization tools will avoid unintentional lying and excessive junky elements.

12.5.1 Advancing Business Intelligence

When thinking about business intelligence (BI), most will immediately think of the well-known BI dashboard. These dashboards are usually reporting portals, though there are dashboard-like tools emerging that do allow iterative exploration and the ability to mash up real-time data with historical data. These capabilities help expand the story that the dashboard information is trying to tell by allowing the testing of various scenarios to drive new conclusions about future business planning. Often, these analytical tools rely on statistical analysis and incorporate all the various analytical categories, including descriptive, diagnostic, predictive, and prescriptive analysis.

Of the many advantages to these immersive dashboards, utility stakeholders can quickly take stock of important smart grid systems, including the ability to measure system performance over time; make decisions that maximize asset utilization; forecast capacity; detect nontechnical losses; and track the effectiveness of demand-side management (DSM) programming, compliance factors, and other metrics. And in spite of the strong interest in analytics-driven BI, the movement toward widespread use of these tools has been incremental at best. Emerging dashboards and portals are being integrated with standard business applications, working to enhance traditional capabilities with visualization tools, modeling applications, and advanced rules and business-logic configuration features.

As costs begin to fall for collecting, processing, and storing big data, and the silos within the utility break down, energy providers will begin to see major advances in the domain of analytics-driven BI. In the utility especially, BI data-discovery techniques will by necessity begin to incorporate real-time-streaming event data. The biggest shift is that the users of these systems will essentially be the authors of the intelligence findings, no longer being served precooked versions of events. However, this evolution is completely dependent on the ability of the organization to pull in disparate forms of data, manage them, and iron out any confusion in the utility about how to deliver this data to users. It will take several years to fully transform, but BI is beginning to shift from a reporting-centric IT focus to an analytics-centric user focus.

12.5.2 High-Impact Operations

Grid analytics are comprised of many data classes, including state, event, signal, consumption, and engineering data. As a result, in analyzing network data for smarter operations, many grid components—such as smart meters, distribution assets, sensors, control devices, intelligent electronic devices (IEDs), communications, and application data—must be presented in a way that allows control center operators to understand the source data, the associated analytics, and the comprehended next action. The best way to do this in the operational domain is through the use of metaphorical objects placed within a geographic information system (GIS) and in a topological context, especially when real-time response and high-impact forecasting is required.

The demand for rapid forecasting in the utility is being driven in large part by the explosive growth of cities that are putting huge demands on the existing infrastructure. Thus, there have been significant strides made in efforts to make optimal use of as much of the available data as possible to improve operational outcomes. The utility, like the smart city, must combine existing databases with sensor information to get a reliable view of the current situation at any moment. But, it is the visualization system that elevates utility operations to meet international, regional, and local standards for operation. Shouldered with extensive responsibilities, grid operators must not only be able to understand the current environment; they must accurately anticipate emerging problems to enable an appropriate response.

Consider that for every major blackout in North America over the past 50 years, one phrase is held in common: "We were unaware." And as stochastic generation technologies are enabled by the smart grid and demanded by citizens and regulators, forecast uncertainty and the lack of comprehensive data to anticipate problems will increase the difficulty of making appropriate control decisions. Tools, such as data analytics and discovery-based analytics, will help operators understand not only existing conditions but also potential conditions.

To accomplish this, operations are moving toward the use of state-of-the-art analytics in their ecosystem, augmenting existing reports but primarily integrating powerful visualization capabilities. Through the use of highly intuitive user interfaces, these operations have seen direct benefits for the rapid detection of faults and grid anomalies. As organizational borders begin to dissolve across the utility, grid data will be stored for further historical analyses in traditional data stores for use in data-trending exercises, asset utilization studies, and post-processing of grid-related events for further exploration.

12.5.3 Improving Customer Value

As discussed in previous chapters, the world of customer analytics is beginning to conflate within the utility. Customer analytics applications serve both customers and customer service representatives, and customers certainly expect the utility to have access to the same information they have, even with the same representation. More and more, utilities are turning to the self-service model that Internet-enabled, smartphone-carrying customers demand. When the power goes out, customers want to report outage information, see maps of affected areas, discern the impact of weather on their area, and obtain a reasonable restoration estimation—all from their phone, which often still has a battery charge and available cellular service. Further, utilities rely on advanced analytics and a mashup of various data sources to cull the causes of their nontechnical losses, identify load patterns, and implement demand-response programs for residential and commercial and industrial (C&I) customers alike.

As the barriers begin to fall between the customer and the utility, the utility has an opportunity to interact with consumers on a regular basis, outside of the once-a-month bill touchpoint. For example, the utility can collaborate with municipalities, institutions, and individual consumers to foster energy conservation, communicate about energy use and how it may impact the monthly bill, offer custom-tailored programs and services, and reduce peak demand. While the smart grid enables the technical foundation for the bidirectional flow of power, its digital infrastructure also brings new information paths to both the utility and, more and more, the customer.

It is precisely this use of information, supported by an analytical backbone, that affords the utility the ability to transform itself into a service entity, if it desires. Right now, the utility handles customers only in the context of an account to which a financial transaction is hinged. Accessible information is the lynchpin in the effort to move toward high levels of service and satisfaction that will prevent the utility from eroding and even building trust.

As energy consumers become more sensitive to their need to exert control over their energy use, there will be increasing demand for tools and information that allow them to perform their own analysis. Utilities can tune these systems to help meet their regulatory and business goals as well, by presenting information and interactive interfaces that will help inspire action. Normative feedback is also proving to be a very important energy-efficiency and conservation tool, as there is a measurable response when consumers learn how they compare to their peers, or how they are contributing to greenhouse gas emissions. Utilities are trying many modalities now—in-home displays, paper reports, web portals, smartphone applications, and smart thermostats. At the heart of this market splintering is an unclear path forward for engaging consumers and driving them to action. This will likely shake out further as major industry players are beginning to move into cleantech, acquiring technology, and consolidating efforts.

Customer analytics is also a growing factor in successful customer service operations, including DSM programming. It can be argued that one of the most important assets a utility possesses is its customer data, since an analytics program that aggregates and combines customer data with other sources not only can identify, confirm, and correct customer issues quickly but also can inform all aspects of utility planning. Smart meter analytics, combined with the benefit of increased data measurement and frequency of collection, make it possible to pinpoint service issues, offer targeted products and services, and reduce theft.

Even with smart meters, revenue losses are continuing to plague utilities. Advanced sensor technologies are beginning to emerge that help utilities work their way up the tree to identify nontechnical losses, and they're making the monitoring and managing of this data quite cost-effective when measured against the magnitude of the growing losses. This same data can also support business processes to assist credit-challenged customers, manage orders, pursue collections, provide customer care, and even minimize regulatory risk. All of these features support a consistent revenue stream and improve margin performance through a smart grid. In every case, the data must be analyzed and presented to analysts, operators, and business users to define appropriate action and response.

Advanced analytics with exploratory and immersive tools enable utilities to organize customer data and turn it into actionable intelligence that improves service, controls costs, and enhances responsiveness when presented with events and visual alerts. Over time, the democratization of data will bring with it the power to analyze, and both the customer service representative and the field worker will be empowered to explore data to identify theft, improve service, and decrease restoration time. Plus, as the utility pushes forward to remake itself in the context of the digital grid, overall customer relations will certainly improve.

12.6 Making Sense of It All

Data visualization can help the utility make sense of big data as well as communicate information once it has been analyzed. There is incredible value to be found with big data, and analytics are the way to extract it. But if the patterns cannot be found or understood, they will never carry any significance—that is the key value proposition of visualization: to make it easier to understand. Despite the fact that the grid itself is made up of physical things, much of the information about utility operations are nonphysical, and even the statistical data is abstract in a big data universe. Comprehending how the human translates this abstract information based on the characteristics of vision requires a close adherence to well-researched and well-defined design principles to ensure that the picture really is, as they say, worth a thousand words.

Things to Keep in Mind When Choosing Visualization Tools
Technology is evolving rapidly. Is the tool backed by a mature company?
Has the tool been used in conditions similar to the proposed use in your utility?
What are the complaints and concerns of the current users of the system?
How easy have the existing integrations been in terms of friction and facility of rollout to the users?
Does it run on the platform you have chosen for your analytics project?

Chapter Thirteen

A Partnership for Change

Called the picture of the century, this is the first view of the earth taken from the moon by the Lunar Orbiter I on the 23rd of August in 1966. (*Source:* NASA[1])

13.1 Chapter Goal

In the final chapter, we discuss how important it is that the utility becomes a trusted steward of not only big data but of how it is analyzed and used. This

---

[1] Image retrieved from the public domain at http://www.dvidshub.net/image/700941/lunar-orbiter-moon-earth.

stewardship is one of the vital factors in building a relationship with the very customers that the utility hopes to work in partnership with for a more reliable, optimized, and distributed power delivery system. The utility of the future will rotate upon this axis of mutuality, enabling the cooperation that is required to manage the inexorable change in the energy delivery sector.

13.2 With Big Data Comes Big Responsibility

It was not Spiderman's Uncle Ben who was the first to note that "with great power comes great responsibility,"[2] but Voltaire. Nonetheless, when it comes to the implications of big data collection and analytics, this rumination is chillingly relevant. There are a few things a utility doesn't want to face: public embarrassment, suspicious consumers, and legal burdens. These are not casual concerns. It is highly likely that many organizations that use personally identifiable information (PII) for their business operations will have to contend with reputational damage as a result of their use or handling of big data.[3] As the utility moves forward with its big data plans, dangerous dilemmas are emerging. Now is the time for utility leaders to explore the implications of using big data for decision-making, especially as the industry lurches from a highly regulated model of universal access to one of optimization, powered by advanced analytics.

The ability of big data to influence is profound, and it amplifies the need to assert a system of values, especially in the areas where the use of big data analytics is designed specifically to drive particular business goals. In the utility, responsive-pricing and energy-efficiency technologies, as well as other load management strategies, are clear examples of the use of information to exert certain effects. There is no way around it: As big data analytics progress, utilities will know more and more about individual behaviors that were once deemed private. As efforts continue to evolve, the utility will bring in third-party data, drawing together millions of data points in an effort to innovate. Yet, while big data is ethically neutral, what the utility does with it in carrying out its business is not.

[2]   Adrien Jean, Quentin Beuchot, and Pierre Auguste Miger (1832), *Œuvres de Voltaire, Volume 48*, Lefèvre.
[3]   Frank Buytendijk and Jay Heiser (2013), "Confronting the Privacy and Ethical Risks of Big Data," *Financial Times*. Retrieved February 17, 2014, from http://www.ft.com/cms/s/0/105e30a4-2549-11e3-b349-00144feab7de.html#axzz2tbFLLaUe.

13.2.1 Abandon All Hope, Ye Who Enter Here?

How do we resolve the need to collect data to analyze and understand customer behaviors that will improve efficiency and conservation outcomes while protecting the fundamental right of privacy? In 2012, the White House revealed a blueprint for online protection called The Consumer Privacy Bill of Rights that underscored the problem, asserting the following statement from President Obama: "[I]nnovation is enabled by novel uses of personal information. So, it is incumbent on us to do what we have done throughout history: apply our timeless privacy values to the new technologies and circumstances of our times."[4] Opinions about what "timeless privacy values" are and how they should be codified, implemented, and governed by the utility vary greatly. Yet, with much work to be done, there have been strides forward that help inform a meaningful dialogue about the requirements for consumer protection.

Unfortunately for the utility, it has taken severe smart-meter backlash, including charges of domestic espionage, for many utilities to begin to take a serious look at privacy. It's one thing for a social media site to know enough about its target customer to present an advertisement for purple cowboy boots; it's entirely another to be collecting information that can reveal a home's daily routines, shifts in those routines, and the types of appliances in the home—down to the moment when the jets in the hot tub go on. This information can help utilities become more efficient and enable them to better market to customers; it can also help an insurance company adjust homeowner's insurance rates based on actuarial profiles, assist a court in subpoenaing witnesses to support a legal position, and help a criminal plan a burglary.

On the topic of smart meters, the European Union (EU) stated that smart metering systems, "enable massive collection of personal data which can **track what members of a household do within the privacy of their own homes** [EU emphasis], whether they are away on holiday or at work, if someone uses a specific medical device or a baby-monitor, [or] how they like to spend their free time."[5] And as early as 2010, the US-based National Institute of Standards

---

[4] Danny Weitzner (2012), "We Can't Wait: Obama Administration Calls for a Consumer Privacy Bill of Rights for the Digital Age," blog, the White House. Retrieved February 21, 2014, from http://www.whitehouse.gov/blog/2012/02/23/we-can-t-wait-obama-administration-calls-consumer-privacy-bill-rights-digital-age.

[5] European Data Protection Supervisor (2012), "Smart Meters: Consumer Profiling Will Track Much More Than Energy Consumption If Not Properly Safeguarded, Says the EDPS," press release, European Data Protection Supervisor. Retrieved February 22, 2014, from http://europa.eu/rapid/press-release_EDPS-12-10_en.htm?locale=en.

and Technology (NIST) wrote, "as Smart Grid implementations collect more granular, detailed, and potentially personal information, this information may reveal business activities, manufacturing procedures, and personal activities in a given location. It will therefore be important for utilities to consider establishing privacy practices to protect this information."[6] Positive actions are far more important than good intentions when it comes to privacy—if the data is available, unethical, nefarious, and criminal schemes will be hatched for its use.

It's not just the collection and maintenance of consumption data from smart meters that provide unprecedented channels to contract information about personal behavior. For example, electric vehicle (EV) owners will have information logged about their batteries' characteristics, and the data, time, and location of their last charge. Smart meters also often serve as a gateway to the consumer's home. Through this gateway, the utility is capable of monitoring devices in the household, including washing machines; hot water heaters; lights; heating, ventilation, and air conditioning (HVAC) systems; pool pumps; and the pervasive (and somewhat vague) Internet of Things. The utility, thus, can feasibly selectively signal any enabled devices to alter those appliances' operation. Kevin Ashton described this technological monitoring in 2009 when he coined the "Internet of Things" (IoT) term, unfolding a new vision of computers that knew everything there was to know about our world by quietly collecting data. He said, "We would be able to track and count everything, and greatly reduce waste, loss and cost. We would know when things needed replacing, repairing or recalling, and whether they were fresh or past their best."[7] It is indeed this vision that forward-thinkers in the industry have in mind as the utility's shift from infrastructure to services is pondered and planned.

13.3 Privacy, Not Promises

Privacy and security issues are complex, but for the consumer, there are three key components to privacy protection: consent, data management, and governance. In response to concerns about all three of these factors, many states and countries are working to develop new approaches to meet privacy considerations.

---

[6] National Institute of Standards and Technology (2010), "Guidelines for Smart Grid Cyber Security: Vol. 2, Privacy and the Smart Grid," The Smart Grid Interoperability Panel—Cyber Security Working Group, NIST. Retrieved February 22, 2014, from http://csrc.nist.gov/publications/nistir/ir7628/nistir-7628_vol2.pdf.

[7] Kevin Ashton (2009), "That 'Internet of Things' Thing," *RFID Journal*. Retrieved February 22, 2014, from http://www.rfidjournal.com/articles/view?4986.

13.3.1 Consent

Consumer consent is related to collecting, managing, and disseminating PII and consumption information. Depending on the data, how it's managed and shared drive the need to garner customer consent. And while every jurisdiction and its related regulatory structure have a stake, any position may be overwhelmed by higher governing entities, including national and supranational bodies.

The use of aggregated customer data is one way to control risk by obscuring the customer's identity yet still allow for useful analytical investigation. But, given the powerful capabilities of analytic tools, reverse engineering is far too easy a task. Recall AOL Internet user No. 4417749, who searched for "dog that urinates on everything" and "60 single men." Her data was released in 2006—along with 657,000 other Americans' information—by AOL, supposedly in anonymized form. However, by following the clickstream data, *The New York Times* quite easily located 62-year-old Thelma Arnold in Lilburn, Georgia.[8] This was a rude surprise.

Some regulators have enacted rules that require aggregated data and prohibit the release of aggregated data (without consent) unless there are at least 15 consumers in the group. Ms. Arnold would likely express that this is entirely insufficient.

The EU mandates that consent consist of affirmative acts (a specific opt-in) that are unambiguous and specific, with a particular focus on what can be done with user data without the benefit of consent. Along these lines of thinking, the EU recommends that the legal basis for choice include everything that is not the following:

> [F]reely given, specific, informed and explicit consent would be required for all processing that goes beyond . . . (i) the provision of energy, (ii) the billing thereof, (iii) detection of fraud consisting of unpaid use of the energy provided, and (iv) preparation of aggregated data necessary for energy-efficient maintenance of the grid.[9]

A further implication of consent in the EU's recommendation is that customers know not only how their data is to be used but also the logic of any

[8] Michael Barbaro (2006), "A Face Is Exposed for AOL Searcher No. 4417749," *The New York Times*. Retrieved February 23, 2014, from http://www.nytimes.com/2006/08/09/technology/09aol.html?pagewanted=all&_r=.

[9] Article 29 Data Protection Working Party (2013), "Opinion 04/2013 on the DPIA Template Prepared by Expert Group 2 of the Commission's Smart Grid Task Force." Retrieved February 23, 2014, from http://ec.europa.eu/justice/data-protection/article-29/documentation/opinion-recommendation/files/2013/wp205_en.pdf

algorithms used in the data analytics performed on their individual household profile, as well as what events might cause them to be subject to disconnection or further scrutiny.

In the United States, privacy principles are part of the Bill of Rights, stated explicitly in the Fourth Amendment, which prevents unreasonable search and seizure. Smart meter data specifically is covered by statutory protections such as the Electronic Communications Privacy Act (ECPA), the Stored Communications Act (SCA), the Federal Trade Commission Act (FTC Act), and the Privacy Act, as well as state-level rules and regulations. In Canada, the Ontario Office of the Information and Privacy Commissioner (IPC), Dr. Ann Cavoukian, has taken a lead with her framework Privacy by Design (PbD), which calls for embedded privacy and data protection throughout the entire life cycle of technology.

Utilities, by virtue of smart meter consumption data collection, now have access to detailed information on household activities. This near-real-time metering can and has been interpreted as a human-rights issue. Article 17 from the Universal Declaration of Human Rights, collectively signed and ratified by most nations, says,

1. No one shall be subjected to arbitrary or unlawful interference with his privacy, family, home or correspondence, nor to unlawful attacks on his honour and reputation.
2. Everyone has the right to the protection of the law against such interference or attacks.[10]

The European Convention on Human Rights (1950), Article 8: Right to Respect for Private and Family Life, more explicitly states:

1. Everyone has the right to respect for his private and family life, his home and his correspondence.
2. There shall be no interference by a public authority with the exercise of this right except such as is in accordance with the law and is necessary in a democratic society in the interests of national security, public safety or the economic well-being of the country, for the prevention of disorder or crime, for the protection of health or morals, or for the protection of the rights and freedoms of others.[11]

---

[10] General Assembly of the United Nations (1966), "International Covenant on Civil and Political Rights." Retrieved February 23, 2014, from http://www.ohchr.org/en/professionalinterest/pages/ccpr.aspx.

[11] European Convention on Human Rights and Its Five Protocols (1950). Retrieved February 23, 2014, from http://www.hri.org/docs/ECHR50.html.

Even to the layperson, it is clear that these articles present us with a test, of sorts. Does our big data analytics project interfere with customer privacy? If so, is the infringement activity in accordance with the law? And further, for European nation states, does the infringement serve any of the specific interests of society mentioned in Article 8 of the European Convention on Human Rights? And is the infringement a necessary component in the maintenance of a democratic society?

While we pragmatically speak about fair information practices, decades ago, it was agreed that privacy is a human right, and data privacy can be no exception. In fact, the Dutch Parliament rejected a mandatory rollout of smart meters, in part because it *might* have violated Article 8 of the European Convention on Human Rights. Since then, the development of data privacy policies, protection policies, and governance codes has been taken much more seriously. Clearly, consent is a key function in protecting our universal human rights, and if so much confusion has been wrought over smart meter data itself, the problem will only compound as utilities seek to mash up many data points from various utility and third-party systems. Building in the appropriate protections from the beginning will not only ensure data protection but also provide a framework to support the continuing innovation of data-collecting technology and systems that the utility will want to analyze for further value and return on investment (ROI).

13.3.2 Data Management

Public policy cannot adequately address problems and concerns about privary and security; instead, solutions must be embedded in the technical design. This is called Privacy by Design (www.privacybydesign.ca), which Dr. Cavoukian developed in the 1990s to address the burgeoning systemic effects of information and communications technology (ICT) and large-scale networked systems. Her prescient work was translated to the domain of the smart grid and encompasses not only business practices but also the critical role of information systems and the physical design of the computing and network infrastructure itself, and it continues to inform the general practice of big data analytics.

Along the continuum of data management from collection to secure transfer and ultimately to storage, the most significant risk to the continued assurance of data privacy is indefinite storage. If they even exist, retention policies very rarely line up with the original purpose of the data collected. They should, though, as indefinite storage simply increases the risk of data breach. In the software engineering environment, there is a classic acknowledgment that the bigger the body of source code grows—and the more engineers who touch the code for maintenance and enhancement—the more brittle, defect-ridden, and more easily exploitable the applications of that code base become. The same principle holds true within the domain of data management: As the volumes of data grow—and the information technology environment becomes more complex—the difficulties of securing that data mount.

Utilities must determine early in the process of building a smart grid data analytics platform what will be the sensitivity of the various types of data that are flowing into the utility, especially as new products and services are developed. This is a very difficult task. Without a lot of clarity about how the utility of the future will evolve, the single greatest cost is perceived to be storage, as volumes of data are put away for a rainy day. This, however, will make governance and compliance nearly impossible. Big data management must include the ability to audit data for its adherence to existing and emerging policies, and part of that auditing scheme will include ensuring compliance with data retention requirements as well as data destruction requirements. In some localities, this will be complex indeed, as data management strategies will be required to adhere to the same expectations for privacy as those that the consumer has specifically agreed to.

13.3.3 Governance

Data governance is a discipline, not a discrete practice. It is the role of data stewards or custodians to ensure that data is handled correctly through a system of processes and methods. However, it's a bit of a dirty little secret in the big data industry (especially heavily regulated industries) that, although analyzing big data holds incredible advantages for almost any organization, the struggle to provide effective governance and embrace privacy regulations is crushing.

It seems unfair that regulated entities must work to enforce required governance while faster-moving players in the data world—including Google, which acquired Nest Labs in 2014 for USD $3.2 billion—rush into the energy sphere.[12] Though many utilities and their regulators were hand-wringing over how to ensure proper data-handling practices to put customers at ease, Nest Labs charged forward by offering a successful service model for home coordination, starting with a smart thermostat. The controllable thermostat has long been the purview of utilities, but it is probable that the advantage the utility had with the smart meter as a gateway to the home has now been severely diminished. There are a multitude of factors at play, but it is surely a hint of how the expectations

---

[12] Rolfe Winkler and Daisuke Wakabayashi (2014), "Google to Buy Nest Labs for $3.2 Billion," *The Wall Street Journal*. Retrieved February 23, 2014, from http://online.wsj.com/news/articles/SB10001424052702303595404579318952802236612.

of data governance issues in the publicly accountable utility industry will stall electricity providers' ability to compete. If regulatory bodies endeavor toward privacy enforcement and protection (and they will), it will be continue to be a formidable challenge for any utility to compete with private-industry players to provide data-enabled services to energy consumers.

Surely, it is completely reasonable for people to have confidence that their personal data will be private and protected until destruction; although the issue of whether data can ever be eradicated makes this even more of a quagmire. The burden of data governance is the question, and it's sure to cause dustups for years to come, especially as regulated utilities work to remake their business model. It's not as if the Googles of the world don't have data governance, as they surely do. The ability to properly handle data is the foundation of their very business. They are just, thus far, not expected to be transparent about how they implement their governance initiatives, simply as a matter of competitive advantage.

Data governance will always be an important concern for energy providers in their big data efforts, and as such, they will be required to trace data, remediate failures, and audit and report on their activities. In fact, in the early years of implementing big data analytics programs, it will be much too easy to mishandle data either inside the utility or with partners because of the silos within the utility organization. Establishing a data steward early in the process is an important aspect of rolling out an enterprisewide data analytics platform, especially as regulatory requirements quickly change and emerge.

13.4 Privacy Enhancement

Bruce Schneier, an internationally renowned security technologist, discusses "enabling the trust society needs to thrive" in provoking a move toward tackling the complex issues of cooperation in a world where digital mediation rules. He treats this issue in his book Liars and Outliers, where he characterizes the dilemma by saying, "In the absence of personal relationships, we have no choice but to substitute security for trust, compliance for trustworthiness. This progression has enabled society to scale to unprecedented complexity, but has also permitted massive global failures."[13] If the utility is to be a useful entity to society in the future, issues of trust and security are critical to develop with communities and customers.

[13] Bruce Schneier (2012), *Liars and Outliers: Enabling the Trust That Society Needs to Thrive*, John Wiley & Sons Inc. Retrieved February 23, 2014, from https://www.schneier.com/book-lo.html.

The ways that lawmakers, regulators, companies, technologists, and citizens have thought about security and privacy are changing. In particular, the manners in which we enable consent and manage and use data are changing from traditional approaches.

13.4.1 Enabling Consent

We're all familiar with the long and onerous Terms and Conditions statements that few people actually read as we rush through the checkboxes (often using our browser's autofill functionality) to get to the desired application or service as soon as possible. But that doesn't mean consumers don't care about privacy: When Facebook or Google asks whether we would trade our expectations of privacy for free services, many of us agree. There is a very immediate value-add, and it's a clear choice on the part of the consumer to give a little to get a little. In the realm of power delivery, there is no clear benefit, and many become very irritated by the idea of having their behavior tracked, especially by a commodity product provider. This is a sensitive point, particularly when it becomes clear to consumers that utilities are now in the position to mine sensitive data—and that they can do so for any reason, even if it is not absolutely necessary to keeping the wires up and the electrons flowing. It's a dangerous position for the utility to take: that customer data is simply a corporate asset.

In the utility business, opt-in and opt-out as functional mechanisms for consent are complicated. Driven largely by customer backlash to smart meters—for reasons including fear of "utility spy programs"—an increasing number of jurisdictions offer opt-out programs, through the choice to keep an analog meter, sometimes with an accompanying fee (referred to by some consumer advocates as "extortion"). And, in regions where mandatory smart meters have been rejected as a matter of policy, voluntary participation by explicit opt-in is now the rule. While many would argue that these cases are clear demonstrations of the abysmal level of trust for the utility, it is worth considering that utilities simply severely miscalculated the nature of the customer relationship in a world where the utility delivers a commodity product and where there is no real qualitative differentiation among providers (in competitive markets). Instead, cost is the key sensitivity to consumers who, until very recently with the emergence of climate-change concerns, didn't care if their electricity was delivered by a peasant woman carrying their electrons in a woven basket. If a proactive effort is not made from the outset to put customer concerns first, the customer backlash to the utility implementation of smart meters is just a foreshadowing of the public's unwillingness to accept an acknowledged big data analytics program.

Notice and consent have always been the cornerstone of personal privacy, but in the world of big data, this is not sustainable, both in terms of applying such a position to the vast volumes or data and in terms of the incredible burden this places on the individual. Plus, in many cases, the horse is already out of the barn: There's no going back to provide consent for all the data that has already been collected, used, and resides in the dark corners of the enterprise. Going forward, ensuring privacy protection is squarely upon the organizations that use the data. In this light, the industry must recognize the importance of fitting data collection programs and remain accountable to how the data is processed and used. The concepts of explicit consent will likely morph and shift in the next decade of big data and be subsumed by a model that instead focuses on regulating the acceptable use of data, rather than the data itself.

13.4.2 Data Minimization

Like consent, the role of data minimization is changing with the pervasiveness of big data in collection, processing, and new storage approaches. For many years, at least in the US states and EU nations, minimization has especially targeted the source of the data, with regulations for the "reasonable" collection of data. Big data analytics is creating a shift in the application of this fundamental right because lots of data is required to extract meaningful value. Thus, there is a transference of efforts for data minimization at the point of collection to the point of use because traditional data minimization approaches break big data.

New frameworks to drive privacy and data protection principles acknowledge the difficulty in data minimization. The emerging argument asserts that the expectation for privacy must be measured against societal value, thereby sublimating the need for minimization (and potentially even some forms of consent). Privacy advocates Omer Tene and Jules Polonetsky suggest the use of a risk matrix, stating the following in the discussion "The Privacy Paradox" in a *Stanford Law Review* issue: "A coherent framework [takes] into account the value of different uses of data against the potential risks to individual autonomy and privacy. Where the benefits of prospective data use clearly outweigh privacy risks, the legitimacy of processing should be assumed even if individuals decline to consent."[14]

---

[14] Jules Polonetsky and Omer Tene (2012), "Privacy in the Age of Big Data: A Time for Big Decisions," *Stanford Law Review*, online, vol. 64, no. 63. Retrieved from http://www.stanfordlawreview.org/online/privacy-paradox/big-data.

13.4.3 The Role of Metadata

A critical concern related to big data privacy is that of data quality and accuracy, especially when that data includes PII, because inaccurate data can lead to a negative effect through flawed results. This can be especially troublesome if these flaws concern a particular consumer or class of consumers who may be targeted by a certain analysis. The various ways in which data quality can break down leads to an important guiding principle for data privacy: The data that's collected should be directly relevant to its intended purpose of use.

To assure data is as accurate as possible, a two-pronged approach is required: data quality requirements and assessment. However, with the absolute volume and pace of information flowing into the utility, it is not feasible to examine every byte of data. Instead, creating a comprehensive system of descriptive metadata proves fruitful. With such a system in place, the challenge shifts to creating a description of the underlying data that can be used as a frontline defense.

The importance of metadata has long been known for its essentialness to data management, and its value is even greater in the era of big data. In fact, the importance of metadata is emerging in surprising ways, as it is even possible to find relationships within the metadata itself to surface system issues or information that analysts are especially looking for and then digging deeper into the actual data.

Metadata is crucial to the process of finding inaccurate data in scores of information, but it can even guide analytic functions by helping to understand data properties. Additionally, metadata provides the foundation of ongoing data-quality monitoring.

13.5 The Utility of the Future Is a Good Partner

The utility is transforming from a one-way network to one of distributed energy resources (DERs), and this shift begs the question of whether the business models for the industry must change. While advanced analytics may be the key enabling technology, it is really the customers themselves who will play the most significant role in producing and managing energy that enables an entirely new economic system. In the utility of the future, the customers are not ratepayers; they are partners. And successful partnerships are not made by exploitation, nor just by being a "good guy," but by building a form of social equity that is based in mutually increased value.

Utilities have gone far in their implementations of smart meters and are working to outfit their grids with further advanced technologies, especially sensors and control devices. This advancement is key to enabling a grid that will remain stable and resilient in an ecosystem that includes a growing amount of intermittent generation sources. There is no question that performing big data analytics on the influx of varied data will be the key to discovering and implementing grid optimizations and introducing efficiencies that can fundamentally move the utility business model forward. In the course of its evolution, the electricity industry is likely to find itself deep within an interconnected web of a variety of participatory business models, where customers of all types interface directly with the distribution grid. And there are many technical challenges that stand between today's technology infrastructure and the utility of the future.

The Rocky Mountain Institute describes how services and control interfaces are now shared among electricity distribution systems that are owned and operated by utilities and customers, utilities, and third parties: "The services provided by distributed resources can include energy and capacity, as well as ancillary services such as the provision of reserves, black-start capability, reactive power, and voltage control."[15] It is not only the continuing demand for increased resiliency and reliability that is driving this shift in the context of extreme conditions that necessitates this change, but the increase in affordable solar photovoltaic (PV) solutions and the growing cultural shift toward the integration of small, local power generation.

It is no surprise to industry observers that the industry is ripe for bypass, largely because of priorities for cleaner generation resources and rate structures that repeatedly fail to meet the value of services to customers. Utilities can fight this trend and spend future years turning to regulators over and over to raise rates on their remaining customers, or they can see the situation as an opportunity to adapt to a new way of doing business. Because, even as these utilities lose their customers to cheaper sources, they will still be required to offer an interconnected grid that is flexible and predictable, alongside a growing penetration of intermittent supplies. Why squander this time of dramatic change to not just survive, but thrive?

Thriving requires investment in core competencies. Certainly, the utility will be asked to make major investments in smarter technologies that help meet the needs of this rapidly changing energy ecosystem. There are several opportunities, and each move will require a step function in information technology improvement and the ability to quickly utilize a variety of big data sources to support these models. In fact, without a fully realized analytics program, the utility will be poorly positioned to innovate at all toward capturing new profit-making opportunities. A not-so-distant fortuitous opportunity includes supporting automatic price signals that help manage supplies across the distribution

---

[15] Rocky Mountain Institute (2013), "New Business Models for the Distribution Edge," Retrieved from http://www.rmi.org/PDF_eLab_New_Business_Models_Report.

system and fostering an economic system that fairly compensates customers for their services but charges them appropriately for power and other amenities they may receive from across the grid.

New business models are enabled by smart grid data analytics. Utilities that wishes to increase their future potential will explore, integrate, and implement new services that create value for electricity customers and the utility alike. Big data analytics is not just a passing phenomenon—it is the lynchpin to fundamentally changing the way utilities operate and interact with their customers. Don't be distracted by the buzzword parade marching across the parking lot: "big data," "big energy," "big value." Instead, focus on the need to rethink and recharge fundamental business models and recognize that even a century of reliable power delivery guarantees nothing for the future. The optimized utility can only be built on the power of data analytics and subsequent high-value action that changes the way that business gets done.

Glossary

Chapter 1

3V	volume, velocity, and variety
ARRA	American Recovery and Reinvestment Act of 2009
DER	distributed energy resource
DMS	distribution management system
EIA	US Energy Information Administration
EPRI	Electric Power Research Institute
ETL	extract, transform, and load
IT	information technology
MDMS	meter data management system
OMS	outage management system
OT	operations technology
ROI	return on investment
SCADA	supervisory control and data acquisition

Chapter 2

AaaS	Analytics as a Service
API	application programming interface
COTS	commercial off-the-shelf
DAMA	Data Management International
HTTP	Hypertext Transfer Protocol
IaaS	Infrastructure as a Service
IT	information technology
MDM	master data management
PUBSUB	publish–subscribe messaging pattern
ROI	return on investment

SOA	service-oriented architecture
SSOD	single source of data
SSOT	single source of the truth

Chapter 3

C&I	commercial and industrial
FIFO	first-in, first-out
IEEE	Institute of Electrical and Electronics Engineers
IoT	Internet of Things
NIST	National Institute of Standards and Technology
ROI	return on investment
SOA	service-oriented architecture
T&D	transmission and distribution
TAFIM	Technical Architecture Framework for Information Management
TOGAF	The Open Group Architecture Framework

Chapter 4

GIS	geographic information system

Chapter 5

BI	business intelligence
C&I	commercial and industrial
DROMS	Demand Response Optimization and Management System
FERC	Federal Energy Regulatory Commission
FLISR	fault location, isolation, and service restoration
kWh	kilowatt-hour
NILM	nonintrusive load monitoring
MW	megawatt
ROI	return on investment
VAR	volt-ampere reactive

Chapter 6

AMI	advanced metering infrastructure
API	application programming interface
CAISO	California ISO

CIM	Common Information Model
CRN	Cooperative Research Network
DA	distribution automation
DER	distributed energy resource
DMS	distribution management system
FLISR	fault location, isolation, and service restoration
IEC	International Electrotechnical Commission
IEEE	Institute of Electrical and Electronics Engineers
ISO	independent system operator
KPI	key performance indicator
NIST	National Institute of Standards
NRECA	National Rural Electric Cooperative Association
OMS	outage management system
PEV	plug-in electric vehicle
ROI	return on investment
SCADA	Supervisory Control and Data Acquisition

Chapter 7

ATM	automated teller machine
BG&E	Baltimore Gas and Electric
CRM	customer relationship management
GIS	geographic information system
HAN	home area network
HEM	home energy management
HER	home energy report
HVAC	heating, ventilation, and air conditioning
IHD	in-home display
IPO	initial public offering
KPI	key performance indicator
KTLO	keeping the lights on
MTKD	mean time to kitchen drawer
PLC	power-line carrier
ROI	return on investment
SMUD	Sacramento Municipal Utility District

Chapter 8

APT	advanced persistent threat
BPL	broadband over power line
CERTS	Consortium for Electric Reliability Solutions

CIP	critical infrastructure protection
CTO	chief technology officer
EEI	Edison Electric Institute
FAA	Federal Aviation Administration
FBI	Federal Bureau of Investigation
GAO	Government Accounting Office
ICS-CERT	Industrial Control Systems Cyber Emergency Response Team
ICT	information and communications technology
IP	Internet Protocol
NASA	National Aeronautics and Space Administration
NCCIP	National Cybersecurity and Critical Infrastructure Protection
NERC	North American Electric Reliability Corporation
PMU	phasor measurement units
NCCoE	National Cybersecurity Center of Excellence
SANS	System Administration, Networking, and Security Institute
SCADA	supervisory control and data acquisition
SPAWAR	Space and Naval Warfare Systems Command (US Navy)
SQL	Structured Query Language
UN	United Nations
USD	US dollars

Chapter 9

CPU	central processing unit
DRMS	demand-response management system
DSM	demand-side management
DMS	distribution management system
DER	distributed energy resource
EDI	electronic data interchange
XML	Extensible Markup Language
ETL	extract, transform, load
FIFO	first-in, first-out
GIS	geographic information system
IED	intelligent electronic device
IP	Internet Protocol
IoT	Internet of Things
MDMS	meter data management system
OMS	outage management system
PMU	phasor measurement unit
PEV	plug-in electric vehicle

| ROI | return on investment |
| volt/VAR | voltage/volt-ampere-reactive |

Chapter 10

API	application programming interface
CPU	central processing unit
DAS	direct-attached storage
XML	Extensible Markup Language
ETL	extract, transform, load
FTP	File Transfer Protocol
GIS	geographic information system
HDFS	Hadoop Distributed File System
HPC	high-performance computing
HTTP	Hypertext Transfer Protocol
IMDB	in-memory database
I/O	input/output
IOPS	input/output operations per second
MMDB	main memory database
NAS	network-attached storage
NoSQL	Not Only SQL
NVDIMM	Non-Volatile Dual In-line Memory Module
OODBMS	object-oriented database management system
RDBMS	relational database management system
PMU	phasor measurement unit
SCADA	supervisory control and data acquisition
HTTPS	Secure Hypertext Transfer Protocol
SQL	Structured Query Language
TVA	Tennessee Valley Authority
URL	uniform resource locator

Chapter 11

CEP	complex event processing
DBMS	database management system
DR	demand response
DSMS	data stream management system
HDFS	Hadoop Distributed File System
KPI	key performance indicator
M&V	measurement and validation

PCA	principal components analysis
ROI	return on investment
SCADA	supervisory control and data acquisition
SQL	Structured Query Language

Chapter 12

BI	business intelligence
C&I	commercial and industrial
CAISO	California independent system operator
DSM	demand-side management
GIS	geographic information systems
IED	intelligent electronic device

Chapter 13

AOL	America Online
EPCA	Electronic Communications Privacy Act
EU	European Union
EV	Electric Vehicle
FTC	Federal Trade Commission
HVAC	heating, ventilation, and air conditioning
ICT	information and communications technology
IoT	Internet of Things
IPC	Information and Privacy Commissioner
NASA	National Aeronautics and Space Administration
NIST	National Institute of Standards
PbD	Privacy by Design
PII	personally identifiable information
PV	photovoltaic
ROI	return on investment
SCA	Stored Communications Act
USD	US dollars

Index